KEEPING IT REAL: ALGEBRA I

A Principal and Teacher's Guide for Implementing School-Based STAAR Curricula Standards

Dr. Patricia Hoffman-Miller
Dr. Freddie L. Frazier
Dr. Kelvin Kirby
Mrs. Frances Frazier

Educational Concepts publishing

Cypress, Texas

Keeping it Real: A Principal and Teachers guide for implementing school-based STAAR curricula standards – Algebra I Edition

Copyright © 2014 by (Edusmart Academic Solutions, LLC)
ALL RIGHTS RESERVED. No part of this book may be reproduced or transmitted in any form or by any means without written permission from the author.

ISBN (978-0-9834914-6-0) - perfect bound

Library of Congress Control Number: 2014934826

EDUSMART ACADEMIC SOLUTIONS, LLC
18330 Chapmans Count Road
Cypress, TX. 77433
Phone 713-452-0102 • Fax 281-758-1564

Educational Concepts, LLC
12320 Barker Cypress Road
Suite 600-111
Cypress, TX 77429
info@educationalconcepts4you.com
Phone: (888) 630-6650
Fax: (877) 310-6692

First Edition
First Printing April 2014

Limited Photocopy License
These materials are intended for use only by qualified professionals.
The publisher grants to individual purchasers of this book nonassignable permission to reproduce all student materials for which photocopying permission is specifically granted. This license is limited to you, the individual purchaser, only for personal use or use with individual students. This license does not grant the right to reproduce these materials for resale, redistribution, electronic display, or any other purposes (including but not limited to books, pamphlets, articles, video- or audiotapes, blogs, file-sharing sites, Internet or intranet sites, and handouts or slides for lectures, workshops, or webinars, whether or not a fee is charged). Permission to reproduce these materials for these and any other purposes must be obtained in writing from Educational Concepts, LLC.

Ordering Information:

Quantity Sales: Special discounts are available on quantities purchased by corporations, associations, school districts, and others.

Individual Sales: Educational Concepts publications can be ordered directly from the publisher or from most bookstores.

Orders for College Textbook/Course Adoption Use: Please contact Educational Concepts, LLC or Edusmart Academic Solutions, LLC. directly.

Dedication

This book is dedicated to all the students, parents, teachers, principals and superintendents refusing to accept there is an achievement gap and dedicating their lives to the notion that failure is not an option!

P.H.M.

I certainly do care about measuring educational results. But what is an 'educational result?' The twinkling eyes of my students, together with their heartfelt and beautifully expressed mathematical arguments are all the results I need.

-Keith Devlin

Acknowledgements

Many thanks to Dr. Angela Dickson for her assistance in formatting the standards during the early stages of this project.

This book is dedicated to all who came before me and inspired my thirst for knowledge and excellence. I honor you by sharing the same inspiration and quest for excellence to all my students and colleagues. - P.H.M.

Table of Contents

CHAPTER 1 – The Principal's Role in Understanding, Implementing and Assessing STAAR Curricula Frameworks   1

    Organization of the Manual   2

    How to Implement and Use the Manual   2

    About STAAR Standards and Alignment   3

CHAPTER 2 – What are the STAAR standards for Algebra I?   5

    Functional Relationships   6

    Properties and Attributes of Functions   6

    Linear Functions   7

    Linear Equations and Inequalities   8

    Quadratic and Other Nonlinear Functions   9

CHAPTER 3 – EasyMatch Representative Curricular Framework   11

  Functional Relationships

    A.1: Foundations of Functions   13

  Properties and Attributes of Functions

    A.2: Foundations for Functions   43

    A.3: Foundations for Functions   61

    A.4: Foundations for Functions   69

  Linear Functions

    A.5: Geometry and spatial reasoning   85

    A.6: Linear functions   99

KEEPING IT REAL - ALGEBRA I   GRADE 9

Linear Equations and Inequalities
- A.7: Linear Functions   135
- A.8: Linear Functions   149
- A.9: Quadratic and other nonlinear functions   167
- A.10: Quadratic and other nonlinear functions   189

Quadratic and Other Nonlinear Functions
- A.9: Quadratic and other nonlinear functions   167
- A.10: Quadratic and other nonlinear functions   189
- A.11: Quadratic and other nonlinear fucntions   203

Chapter 1

WELCOME TO THE KEEPING IT REAL SERIES

What is the Principal's Role in Understanding, Implementing and Assessing STAAR Curricula Frameworks?

Understanding, implementing and assessing the newly adopted STAAR curriculum mandated by the Texas Education Agency caused much consternation in districts, schools and classrooms across the state of Texas. The adoption of any new initiative, whether district, state or federal, requires a commitment of school and district time, effort and professional development as a method of ensuring that the entire school community is supportive of the endeavor. Failure to form a cohesive unit during the implementation process may result in under-performing results with severe implications for the entire school.

A strategic and prescriptive approach helps to avoid unnecessarily high levels of stress and anxiety during this critical phase; stress and anxiety that may severely impede your ability as an instructional leader. Unfortunately, the role of a building or district leader dictates that you have a limited amount of time to fully comprehend and guide instruction during this critical phase. Yet, the success or failure of students on your campus depends on your capacity to demonstrate transformational leadership.

As a campus administrator or district curriculum specialist, you are expected to be an expert in all content areas and across all grade levels, irrespective of your teaching expertise prior to assuming an administrative role. In reality, however, we understand that it is impossible to be an expert in all curricula, particularly across numerous grade and content levels.

It was for this purpose that the **"Keeping it Real"** series was created. As former building and district leaders, the authors are more than aware of the requirements associated with the adoption of mandated curricula. Our cogent examples and guidance comes from over thirty years of building and district administrative leadership, teaching and learning. It is through this experience that we present you with tried and true methods successfully utilized in classrooms, buildings and districts throughout the United States and in all content areas. We call our approach "prescriptive" in that it provides teachers, building and district administrators with a prescription for the success of all students. As you explore the following pages, we are sure you will agree with us: **"Keeping it Real"** is designed to provide immediate solutions for 21st century teachers and administrators.

KEEPING IT REAL - ALGEBRA I   GRADE 9

ORGANIZATION OF THE MANUAL

ICON KEY

 Curricular Framework: Lesson Plans, Assessment and Connecting Themes

 Assessing student knowledge

 Integrating Technology

 Writing Across the Curriculum

 Differentiating

As is the case with all **"Keeping it Real"** Curricular Frameworks, Algebra I – Grade Nine is designed to provide an instant snapshot of a representative curricular framework aligned with the most recently released version of the Texas STAAR. The representative curricula, coupled with aligned assessments for each STAAR performance indicator, are immediately ready for implementation at the classroom level, with simultaneous review and evaluation by building principals. There is no need to design new instruction as the instructional content is pre-tested and designed for you!

Why is this important for principals and classroom teachers? It is important in that it provides a customizable lesson plan completely aligned with STAAR expectations. Principals have the capacity to view a "quick shot" link or lesson demonstrating content that should be taught by teachers to assure student learning. Teachers have the ability to download, adopt, modify or replicate the curricular framework immediately without searching for, or developing lessons to assure student learning. Teachers are now able to develop instructional strategies consistent with best practices without focusing on specific content and the attendant rigorous assessments. Each curricular framework incorporates complete lesson plans in addition to appropriate assessments for the standards identified.

Keeping it Real uses the **EasyMatch** Curricula Framework designed by the authors to address continual challenges identified as a result of rigorous academic standards, accountability and measurement of student performance. While the frameworks identified here are consistent with the Texas Education Agency's STAAR, additional materials may be used to supplement student performance as required by district mandates.

How to Implement and Use the Online Material

Icons were developed to identify essential components of the EasyMatch system. Each icon provides the user with an identifiable user interface that describes the nature of the content provided. This system allows the users to modify, adapt, adopt or otherwise use the material to its fullest potential. For example, teachers may decide to use sections of the system that best demonstrates "writing across the curriculum". They have the flexibility to copy, paste and duplicate the material represented by the icon aligned with a particular standard. Concur-

KEEPING IT REAL - ALGEBRA I — GRADE 9

rently, principals may desire to preview lesson content in specific areas to assist in the evaluation of instructional efficacy. Therefore, it would not be necessary to review each EasyMatch icon, but only those icons consistent with instruction identified for that particular standard. This scientific approach focuses not only on the mechanization of instruction but also the assessment of instruction, providing more free time for campus principals and teacher leaders to identify instructional gaps that may be present in individual classrooms and/or specific content areas!

About the "Picture" Icons

The "picture" icons are similar to a tablet icon for an application. Each symbol for this standard is representative of a specific lesson, with assessments, aligned with each **Readiness Standard** or **Supporting Standard.**

For example, an eighth-grade social studies teacher may need assistance in crafting a lesson to address **Standard Three, Government and Citizenship – 3-15:C**. By clicking on, or referring to the Icon associated with this standard, the teacher is able to immediately access a pre-developed lesson plan, including assessments, that best demonstrate best practices for this grade and content level.

Principals, are granted access to the on-line component of EasyMatch allowing ease of access to all representative assignments consistent with purchased grade and content material. Access to EasyMatch Online provides support for principals in the following ways: 1) provision of individual EasyMatch assignments aligned with STAAR summative assessments; 2) formative prescriptive assessments for all purchased grade and content levels; 3) Optional Professional Development to expand professional learning opportunities for teachers and other school leaders; 4) Optional Cultural Proficiency extensions to ensure that all student learning needs are met; 5) Optional access for integrating technology and writing across the curriculum; and Curriculum support where additional prescriptive assistance is required.

This method provides building principals with the ability to readily review content aligned with a particular standard in order to assess instruction and/or prepare for a teacher observation. As a result, the process is beneficial to teachers and principals through the provision of a more prescriptive approach to instruction, providing more time for the evaluation of instruction.

www.keepingitrealeasymatch.com

KEEPING IT REAL - ALGEBRA I   GRADE 9

About STAAR Standards and Alignment

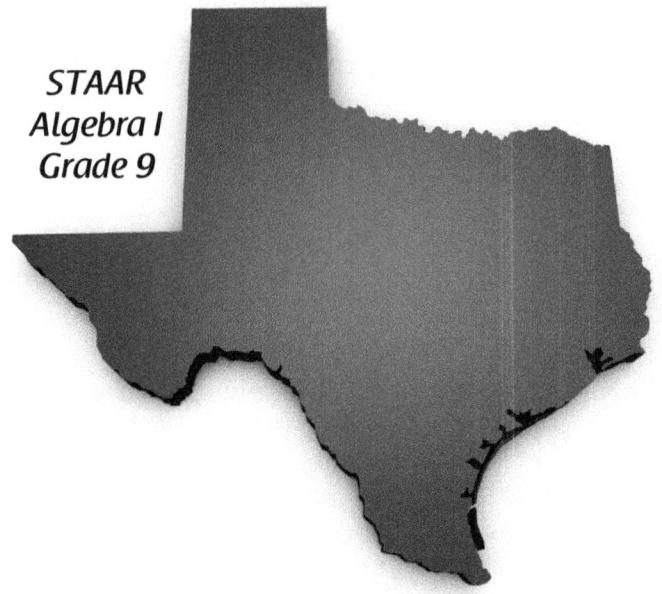

The State of Texas Assessment of Academic Readiness (STAAR) replaced the Texas Assessment of Knowledge and Skills in Spring, 2012. These summative assessments include end of course assessments at the secondary level and grade specific assessments in grades three through eight. All children are expected to perform at the mastery level on this assessment effective the 2012-2013 school year. Curricular frameworks identified in this manual are aligned with **all** readiness and supporting standards contained within the STAAR framework. Exemplary lessons are presented on the following pages consistent with expectations of student performance in **mathematics, grade nine**. Additional curricula are available for each STAAR assessment according to grade and content. All EasyMatch curricular frameworks scaffold learning to ensure horizontal and vertical alignment of subject and grade level requirements.

WHAT ARE THE STAAR STANDARDS FOR ALGEBRA I?

*The State of Texas Assessment of Academic Readiness (STAAR) for Algebra I contain **Five Standards**, commonly known as **Reporting Categories**. These **Reporting Categories**, or **Standards**, provide districts and schools guidelines for student learning.*

The **Reporting Categories** are as follows:

1. **Functional Relationships** - The student will describe functional relationships in a variety of ways.

2. **Properties and Attributes of Functions** - The student will demonstrate an understanding of the properties and attributes of functions.

3. **Linear Functions** - The student will demonstrate an understanding of linear functions.

4. **Linear Equations and Inequalities** - The student will formulate and use linear equations and inequalities.

5. **Quadratic and Other Nonlinear Functions** - The student will demonstrate an understanding of quadratic and other nonlinear functions.

However, STAAR reporting categories contain underlying assumptions pertaining to student learning. These assumptions are framed in mathematical processes and tools where students are required to demonstrate mastery in the application of basic mathematical principles. STAAR assessments for this grade and content level will contain 75% of mathematical processes and tools and are identified along with content standards.

In addition, STAAR reporting categories include content standards which include **supporting** and **readiness** standards representing performance indicators or **benchmarks** to assess student academic readiness. In total, there are **thirty-nine supporting** and **readiness** standards students are required to know and demonstrate mastery of, inclusive of mathematical tools and processes.

KEEPING IT REAL - ALGEBRA I   GRADE 9

The attendant content standards, supporting and readiness standards are listed below:

1. Functional Relationships - The student will describe functional relationships in a variety of ways.

(A.1) Foundations of functions. The student understands that a function represents a dependence of one quantity on another and can be described in a variety of ways. The student is expected to:

(A) Describe independent and dependent quantities in functional relationships; **Supporting Standard**

(B) Gather and record data and use data sets to determine functional relationships between quantities; **Supporting Standard**

(C) Describe functional relationships for given problem situations and write equations or inequalities to answer questions arising from the situations; **Supporting Standard**

(D) Represent relationships among quantities using [concrete] models, tables, graphs, diagrams, verbal descriptions, equations, and inequalities; **Readiness Standard**

(E) Interpret and make decisions, predictions, and critical judgment from functional relationships. **Readiness Standard**

2. Properties and Attributes of Functions - The student will demonstrate an understanding of the properties and attributes of functions.

(A.2) Foundations for functions. The student uses the properties and attributes of functions. The student is expected to:

(A) Identify and sketch the general forms of linear ($y = x$) and quadratic ($y = x^2$) parent functions; **Supporting Standard**

(B) Identify mathematical domains and ranges and determine reasonable domain and range values for given situations, both continuous and discrete; **Readiness Standard**

(C) Interpret situations in terms of given graphs or create situations that fit given graphs; **Supporting Standard**

(D) Collect and organize data, make and interpret scatter plots (including recognizing positive, negative, or no correlation for data approximating linear situations), and model, predict, and make decisions and critical judgments in problem situations. **Readiness Standard**

KEEPING IT REAL - ALGEBRA I — GRADE 9

(A.3) **Foundations for functions.** The student understands how algebra can be used to express generalizations and recognizes and uses the power of symbols to represent situations. The student is expected to:

(A) Use symbols to represent unknowns and variables; **Supporting Standard**

(B) Look for patterns and represent generalizations algebraically. **Supporting Standard**

(A.4) **Foundations for functions.** The student understands the importance of the skills required to manipulate symbols in order to solve problems and uses the necessary algebraic skills required to simplify algebraic expressions and solve equations and inequalities in problem situations. The student is expected to

(A) Find specific function values, simplify polynomial expressions, transform and solve equations, and factor as necessary in problem situations; **Readiness Standard**

(B) Use the commutative, associative, and distributive properties to simplify algebraic expressions; **Supporting Standard**

(C) Connect equation notation with function notation, such as $y = x + 1$ and $f(x) = x + 1$. **Supporting Standard**

3. Linear Functions - The student will demonstrate an understanding of linear functions.

(A.5) **Geometry and spatial reasoning.** The student understands that linear functions can be represented in different ways and translates among their various representations. The student is expected to:

(A) Determine whether or not given situations can be represented by linear functions; **Supporting Standard**

(B) Determine the domain and range for linear functions in given situations; **Supporting Standard**

(C) Use, translate, and make connections among algebraic, tabular, graphical, or verbal descriptions of linear functions. **Readiness Standard**

(A.6) **Linear functions.** The student understands the meaning of the slope and intercepts of the graphs of linear functions and zeros of linear functions and interprets and describes the effects of changes in parameters of linear functions in real-world and mathematical situations. The student is expected to:

KEEPING IT REAL - ALGEBRA I   GRADE 9

(A) Develop the concept of slope as rate of change and determine slopes from graphs, tables, and algebraic representations; **Supporting Standard**

(B) Interpret the meaning of slope and intercepts in situations using data, symbolic representations, or graphs; **Readiness Standard**

(C) Investigate, describe, and predict the effects of changes in m and b on the graph of y = mx + b; **Readiness Standard**

(D) Graph and write equations of lines given characteristics such as two points, a point and a slope, or a slope and y-intercept; **Supporting Standard**

(E) Determine the intercepts of the graphs of linear functions and zeros of linear functions from graphs, tables, and algebraic representations; **Supporting Standard**

(F) Interpret and predict the effects of changing slope and y-intercept in applied situations; **Readiness Standard**

(G) Relate direct variation to linear functions and solve problems involving proportional change. **Supporting Standard**

4. Linear Equations and Inequalities - The student will formulate and use linear equations and inequalities.

(A.7) **Linear functions.** The student formulates equations and inequalities based on linear functions, uses a variety of methods to solve them, and analyzes the solutions in terms of the situation. The student is expected to:

(A) Analyze situations involving linear functions and formulate linear equations or inequalities to solve problems; **Supporting Standard**

(B) Investigate methods for solving linear equations and inequalities using [concrete] models, graphs, and the properties of equality, select a method, and solve the equations and inequalities; **Readiness Standard**

(C) Interpret and determine the reasonableness of solutions to linear equations and inequalities. **Supporting Standard**

(A.8) **Linear functions.** The student formulates systems of linear equations from problem situations, uses a variety of methods to solve them, and analyzes the solutions in terms of the situation. The student is expected to:

(A) Analyze situations and formulate systems of linear equations in two unknowns to solve problems; **Supporting Standard**

(B) Solve systems of linear equations using [concrete] models, graphs, tables, and algebraic methods; **Readiness Standard**

(C) Interpret and determine the reasonableness of solutions to systems of linear equations. **Supporting Standard**

---

5. Quadratic and Other Nonlinear Functions: The student will demonstrate an understanding of quadratic and other nonlinear functions.

(A.9) **Quadratic and other nonlinear functions.** The student understands that the graphs of quadratic functions are affected by the parameters of the function and can interpret and describe the effects of changes in the parameters of quadratic functions. The student is expected to:

(A) Determine the domain and range for quadratic functions in given situations; **Supporting Standard**

(B) Investigate, describe, and predict the effects of changes in a on the graph of $y = ax^2 + c$; **Supporting Standard**

(C) Investigate, describe, and predict the effects of changes in c on the graph of $y = ax^2 + c$; **Supporting Standard**

(D) Analyze graphs of quadratic functions and draw conclusions. **Readiness Standard**

(A.10) **Quadratic and other nonlinear functions.** The student understands there is more than one way to solve a quadratic equation and solves them using appropriate methods. The student is expected to:

(A) Solve quadratic equations using [concrete] models, tables, graphs, and algebraic methods; **Readiness Standard**

(B) Make connections among the solutions (roots) of quadratic equations, the zeros of their related functions, and the horizontal intercepts (x-intercepts) of the graph of the function. **Supporting Standard**

(A.11) **Quadratic and other nonlinear functions.** The student understands there are situations modeled by functions that are neither linear nor quadratic and models the situations. The student is expected to:

(A) Use patterns to generate the laws of exponents and apply them in problem-solving situations; **Supporting Standard**

KEEPING IT REAL - ALGEBRA I

(B) Analyze data and represent situations involving inverse variation using [concrete] models, tables, graphs, or algebraic methods; **Supporting Standard**

(C) Analyze data and represent situations involving exponential growth and decay using [concrete] models, tables, graphs, or algebraic methods. **Supporting Standard**

Chapter 3

EasyMatch REPRESENTATIVE CURRICULAR FRAMEWORK

The EASYMATCH Representative Assignment provides you with:
1) Teacher direction
2) Teacher Materials
3) Lesson Plan
4) Assessment of Learning
5) Suggestions for integrating technology, writing across the curriculum and focusing on special needs learners.

Teachers and principals are often confronted with determining lessons that are appropriate for all learners, addressing what all students should know and be able to master while ensuring instructional standards that are culturally responsive and consistent with reporting and/or readiness standards as prescribed by STAAR. In this context, the central focus of student-centered learning is lost, leading to a "teach to the test" modality, and resulting in teacher burnout and student disaffection. EasyMatch curricular frameworks obviate the need to continually search for lessons aligned with summative assessments, as the representative assignments are identified for you! Teachers are thereby allowed to focus on pedagogy by adopting, modifying or identifying EasyMatch assignments aligned with STAAR. As a Principal, you are no longer required to be a content expert in all areas of instruction. The on-line support system delineates not only the required lesson for specific reporting and readiness categories, it also provides support for embedded professional development and the identification of data assessment strategies designed to focus on student, class and content strengths and weaknesses.

www.keepingitrealeasymatch.com

KEEPING IT REAL - ALGEBRA I   GRADE 9

> *Reporting Category 1*
>
> ## EasyMatch A.1 (A)
>
> **STAAR – Algebra I: Functional Relationships**
> **The student will describe functional relationships in a variety of ways.**
>
> **A.1 – Foundations for functions.** The student understands that a function represents a dependence of one quantity on another and can be described in a variety of ways.
>
> A.1(A) – Describe independent and dependent quantities in functional relationships; **Supporting Standard**

STANDARD AND LESSON OVERVIEW
TEACHER DIRECTIONS

A.1(A) - Describe independent and dependent quantities in functional relationships; **Supporting Standard**

Functions

The statement that **f** is a function means that **f** is a set of ordered pairs (x, f (x)), spoken as "x and f of x," or (x, y) such that **no two ordered pairs have the same first element.** A function has two components which are called the Domain and the Range. The **Domain** is the set of all first elements (**x-values**) and the **Range** is the set of all second elements (**y-values**). For the function (set of ordered pairs) below, find the Domain and the Range:

Function: { (2, 3), (4, 5), (-2, 3), (-4, 5) }

Domain: {2, 4, -2, -4}   Range: {3, 5}

Which of the following sets of ordered pairs are functions?

    a. { (2, 3) (6, 5) (3, 4) (5, 6) }   Y / N
    b. { (2, 3) (2, -3) (4, 5) (6, 2) }   Y / N
    c. { (3, 4) (6, 2) (4, 3) (-3, 2) }   Y / N

The **Domain** of the function is the called the **Independent Variable (x)** and the **Range** of the function is called the **Dependent Variable (y)**. The values for the function are identified **(domain – inputs – x-values)** and then determine the results **(range – outputs – y- values)** for the function. For the function y = x; if x = 1, then y = 1; if x = 2, then y = 2; if x = 3, then y = 3; and if x = 4, then y = 4. Thus, the function y = x consists of the order pairs (0, 0), (1, 1), (2, 2), (3, 3), (4, 4)... and also extends in the direction of (-1, -1), (-2, -2), (-3, -3) and (-4, -4).... .

KEEPING IT REAL - ALGEBRA I

A.1(A)

The function y = x is the **Identity Function** because for each ordered pair, x and y are the same. A plot of the Identity Function is illustrated below in **Figure A.1.1**.

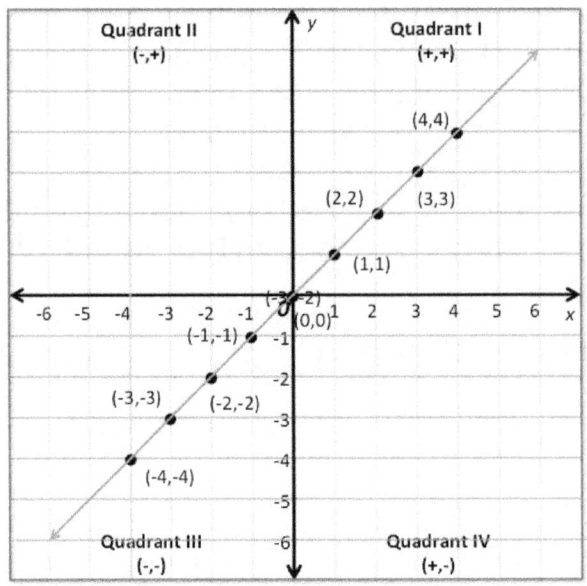

Figure A.1.1 - The Identity Function

www.keepingitrealeasymatch.com

KEEPING IT REAL - ALGEBRA I   A.1(A)

Student Materials

Objectives for A.1 (A) – Functional Relationships

1. Kinko's charges a fixed amount per photocopy and gives 15% discount off the total cost of the photocopies. The total cost is a function of the number of photocopies made. What is the independent quantity in the bargain?
   _____
   _____

2. When pumping gas, the total cost depends on the number of gallons pumped. Identify the independent and dependent variables
   _____
   _____

3. For the table below, determine whether the pairing is a function.

   a.

Input	0	1	8	2
Output	5	4	3	10

   b.

Input	0	0	1	5
Output	10	3	7	15

4. Write an equation which satisfies the function rule that the output is 5 times more than the input. Develop a table with five inputs to demonstrate relationship of the function.

   a. Equation: _____

   b. Develop a table:

KEEPING IT REAL - ALGEBRA I

A.1(A)

5. You and your friends are going to a movie. You agreed to purchase the tickets which are $10 each. You only have $50 to spend on movie tickets. Write the amount, in dollars, you spend as a function of the number of tickets you purchase. Identify the independent and dependent variables, along with the domain and the range.

_____

_____

_____

_____

_____

6. Write a function rule or equation for the ticket purchase situation described in problem 5 above. In your equation, let n represent the number of tickets you purchase and T represent the total dollars spent. Make a table which represents the relationship of the number of tickets purchased and the total dollars spent.

_____

_____

_____

_____

_____

_____

Insert Table Below:

KEEPING IT REAL - ALGEBRA I

A.1(A)

7. Make a table for the function y = x + 5, with a domain of 15, 17, 19, and 21. Identify the range of the function in the table.

Insert Table Below:

8. A function may be represented by a mapping diagram as illustrated below, where an input is paired with an output. Determine whether the pairing is a function. Identify the domain and the range for the function.

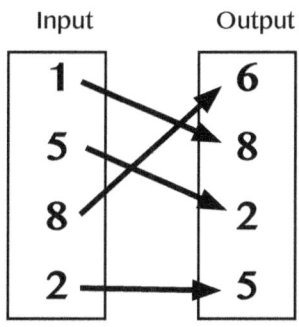

_____
_____
_____

9. Identify the range for the following functions:

    a.   y = 2x + 5   b. y = 3x − 2

    Domain: 1, 3, 4, 5   Domain: 3, 6, 7, 8

_____
_____

KEEPING IT REAL - ALGEBRA I A.1(A)

10. Write a rule or equation for the function below:

Input, x	0	1	2	3
Output, y	5	6	7	8

_____
_____
_____

> # EasyMatch A.1 (B)
>
> **STAAR – Algebra I: Functional Relationships – the student will describe functional relationships in a variety of ways.**
>
> **A.1 – Foundations for functions.** The student understands that a function represents a dependence of one quantity on another and can be described in a variety of ways.
>
> A.1 (B) – Gather and record data and use data sets to determine functional relationships between quantities; **Supporting Standard**

STANDARD AND LESSON OVERVIEW
TEACHER DIRECTIONS

A.1 (B) – Gather and record data and use data sets to determine functional relationships between quantities; **Supporting Standard**

Points

If one can count right, left, up and down, one can plot points. Below in **Figure A.1.2** is a graph with horizontal and vertical number lines. The x-axis is horizontal and shown going left and right. The y-axis is vertical and shown as going up and down. The numbers in the x and y directions are referred to as the x-coordinate and the y-coordinate, respectively. The location where the x-axis intersects or crosses the y-axis is called the origin, which is point (0, 0). At the origin, x = 0 and y = 0. To plot the point (2, 4), where x= 2 and y = 4, start at the origin and count 2 units to the right on the x-axis and stop. From that point, count four units straight up and stop and the point on the graph is (2, 4). To plot the point (-3, 4), start at the origin and count 3 units to the left and stop. Next, count 4 units up and stop. The point on the graph is (-3, 4). Point (2, 4) is in Quadrant I, where x is positive and y is positive. Point (-3, 4) is in Quadrant II, where x is negative and y is positive.

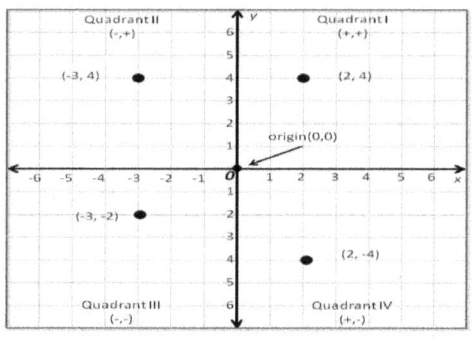

Figure A.1.2

KEEPING IT REAL - ALGEBRA I   A.1(B)

Plotting the point (-3, -2), is located in Quadrant III where x is negative (left 3 units) and y is negative (down 2 units). In plotting the point (2, -4), the point is in Quadrant IV where x is positive and y is negative. A **scatter plot** is a type of display for paired data, where each data pair is plotted as a point on a graph – used to represent sets of data.

In **Figure A.1.3**, plotting the point (0, 2), x = 0 and there is no count to the right or left. The only count is straight up 2 units and the point (0, 2) is located on the y-axis where x = 0 and y = 2. For the point (2, 0), from the origin count 2 units to the right on the x-axis and there is no count up or down on the y-axis because y = 0. Thus, the point (2, 0) is on the x-axis where x = 2 and y = 0.

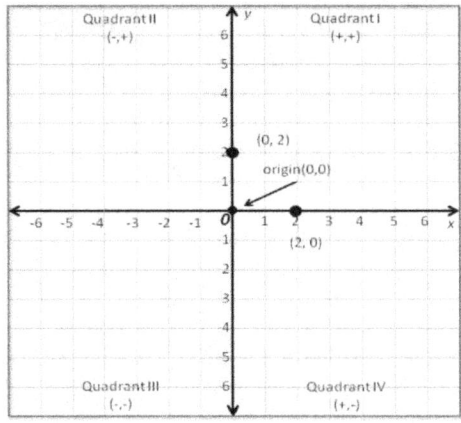

Figure A.1.3

In general terms, a point is a location represented by an ordered pair. In an ordered pair, x is the first coordinate and y is the second coordinate in that order (x, y).

 INTEGRATING TECHNOLOGY    CULTURAL COMPETENCY

 PROJECT-BASED LEARNING    WRITING ACROSS THE CURRICULUM

www.keepingitrealeasymatch.com

KEEPING IT REAL - ALGEBRA I   A.1(B)

Student Materials

Objectives for A.1 (B) – Functional Relationships

1. Using the following data sets, determine the functional relationship and write an equation to represent the function. Identify the domain and the range and specify the independent and dependent variables.

x	y
0	4
1	5
2	6
3	7

_____
_____
_____
_____
_____
_____

2. Make a graph of the function $y = x + 3$. Identify the domain and the range.

Insert graph below:

KEEPING IT REAL - ALGEBRA I   A.1(B)

Describe the relationship between x and y below and graph the functional relationship.

x	y
2	4
4	8
6	12
8	16

_____
_____
_____
_____

Insert Graph Below:

3. Based on the definition of a function, draw a graph which does not represent y as a function of x.

Insert Graph Below:

KEEPING IT REAL - ALGEBRA I

A.1(B)

4. Illustrate a table and mapping diagram for the function y = -2x +3. You select the domain for the functional relationship.

KEEPING IT REAL - ALGEBRA I

A.1(B)

> # EasyMatch A.1 (C)
>
> **STAAR – Mathematics: Numbers, Operations and Quantitative Reasoning – The Student will demonstrate an understanding of numbers, operations and quantitative reasoning**
>
> **A.1 – Foundations for functions.** The student uses place value to communicate about increasingly large whole numbers in verbal and written form, including money.
>
> A.1 (C) – Describe functional relationships for given problem situations and write equations or inequalities to answer questions arising from the situations;
> **Supporting Standard**

STANDARD AND LESSON OVERVIEW
TEACHER DIRECTIONS

A.1 (C) – Describe functional relationships for given problem situations and write equations or inequalities to answer questions arising from the situations;
Supporting Standard

Numbers, Operations and Quantitative Reasoning

In describing functional relationships among quantities, verbal phrases or sentences are translated into expressions such as **equations or inequalities** to answer questions which may arise from the relationship or situation. Some **Definitions** are necessary: An **equation** is a mathematical sentence formed by placing the equal symbol (=) between two expressions or variables. An **inequality** is a mathematical sentence formed by placing one of the symbols $<, >, \leq,$ or $\geq$ between two expressions or variables. An **open sentence** is an equation or an inequality that contains an algebraic expression.

Numbers and operations for functional relationships can be presented in a variety of ways. There are three very important mathematical properties to introduce which will provide consistency and structure to solving functional relationships expressed as **equations or inequalities**. The three properties are the **Additive Inverse, Multiplicative Inverse** and **Distributive Property.**

The **Additive Inverse** is introduced as a one-step equation involving addition. Rational numbers will be included to reinforce concepts introduced in prior mathematics courses.

KEEPING IT REAL - ALGEBRA I A.1(C)

Definition: Additive Inverse property states that if a is a real number then there exist a negative a (-a) such that a + -a = 0. The **opposite of a is its additive inverse (-a)**, such that the sum of a number a and its opposite is 0.

Example 1:   Notes

$x - \dfrac{2}{3} = \dfrac{-3}{5}$ The objective is to get x by itself on the left side of the = sign

$x - \dfrac{2}{3} + \dfrac{2}{3} = \dfrac{-3}{5} + \dfrac{2}{3}$ The additive inverse of $\dfrac{-2}{3}$ is $+\dfrac{2}{3}$, added to both sides

$x + 0 = \dfrac{-9 + 10}{15}$ Recall, $\dfrac{a}{b} + \dfrac{c}{d} = \dfrac{ad + cb}{bd}$, bd is a common denominator

$x = \dfrac{1}{15}$ 0 is the identity element of addition thus, x + 0 = x.

Example 2:   Notes

$x + \dfrac{2}{3} = \dfrac{3}{4}$ The objective is to get x by itself on the left side of the = sign

$x + \dfrac{2}{3} - \dfrac{2}{3} = \dfrac{3}{4} - \dfrac{2}{3}$ The additive inverse of $+\dfrac{2}{3}$ is $\dfrac{-2}{3}$, added to both sides

$x + 0 = \dfrac{9 - 8}{12}$ Recall, $\dfrac{a}{b} - \dfrac{c}{d} = \dfrac{ad - cb}{bd}$, bd is a common denominator

$x = \dfrac{1}{12}$ 0 is the identity element of addition thus, x + 0 = x.

Multiplicative Inverse

The **Multiplicative Inverse** involves another one step equation involving multiplication. Again, rational numbers will be included to reinforce previous concepts.

Definitions: The **Coefficient** of x, is the number to the left of x. Consider a value expressed by 2x (2 times x), where 2 is the coefficient of x. It can also be expressed that the coefficient is the number in front of x. The **Multiplicative Inverse** property states that if a is a real number and a is not zero (0), then there exist another number, $\dfrac{1}{a}$, the reciprocal of a, such that $a \times \dfrac{1}{a}$ or $\dfrac{1}{a}(a) = \dfrac{a}{a} = 1$. Zero (0) does not have a multiplicative inverse because there is no number a, such that 0 x a = 1. Since one (1) is the identity element for multiplication, the multiplicative inverse is a very powerful property of mathematics used in solving various types of equations.

KEEPING IT REAL - ALGEBRA I   A.1(C)

Example 1:   **Notes**

$\dfrac{2}{3}x = \dfrac{1}{4}$ Solving for x, the objective is to make the coefficient of x one (1)

$\dfrac{3}{2} \times \dfrac{2}{3}x = \dfrac{1}{4} \times \dfrac{3}{2}$ Multiply both sides by the Multiplicative Inverse of $\dfrac{2}{3}$, which is $\dfrac{3}{2}$

$\dfrac{6}{6}x = \dfrac{3}{8}$ Based on properties of rational numbers $\dfrac{a}{b} \times \dfrac{c}{d} = \dfrac{ac}{bd}, \dfrac{6}{6} = 1$

$x = \dfrac{3}{8}$ Since 1 is the Identify element of multiplication, solved for x

Example 2:   **Notes**

$\dfrac{-2}{5}x = \dfrac{3}{4}$ Solving for x, the objective is to make the coefficient of x one (1)

$\dfrac{-5}{2} \times \dfrac{-2}{5}x = \dfrac{3}{4} \times \dfrac{-5}{2}$ Multiplicative Inverse of $\dfrac{-2}{5}$, is $\dfrac{-5}{2}$, multiplied on both sides

$\dfrac{10}{10}x = \dfrac{-15}{8}$ Based on properties of rational numbers, $\dfrac{10}{10} = 1$

$x = \dfrac{-15}{8}$ Solved for x

Distributive Property

Definitions: The **Distributive Property** of Real Numbers states that if a, b and c are Real Numbers, then $a(b+c) = a \times b + a \times c$. The number a is multiplied or distributed to both b and c. That is,

$2(3+4) = 2 \times 3 + 2 \times 4$
$2(7) = 6 + 8$
$14 = 14$

Inequalities

A **Linear Equation** is an equation of a straight line, with x raised to the power of one and y raised to the power of one. The Domain is all Real Numbers and the Range is all Real Numbers. A linear inequality in two variables, such as x - 2y < 4, is the result of replacing the = sign in a linear equation with <, ≤, > or ≥. A solution of an inequality in two variables x and y is an ordered pair (x, y) that produces a true statement when the values of x and y are substituted into the inequality. Given the inequality 2x + 4y ≥ 12, rewrite the inequality in **Slope**

KEEPING IT REAL - ALGEBRA I   A.1(C)

Intercept form (**y = mx + b**) and graph, where m is the slope of the line.

Example   Notes

$2x + 4y \geq 12$

$-2x + 2x + 4y \geq -2x + 12$ Additive Inverse; add $-2x$ to both sides

$(4y \geq -2x + 12)$ Multiplicative Inverse of 4 is 1, make the coefficient of y one $\frac{1}{4}$

$\frac{1}{4}(4y \geq -2x + 12)$ Distributive Property applied in next step

$y \geq -\frac{1}{2}x + 3$

The line is $y = -\frac{1}{2}x + 3$ and the Slope Intercept form is $y \geq -\frac{1}{2}x + 3$. Therefore, the line is a **solid line** and the $\geq$ requires that the **area above the line** is shaded. The y-intercept (where x = 0) is the point (0, 3) and the slope m is a $-\frac{1}{2}$ or the $\frac{rise}{run}$.

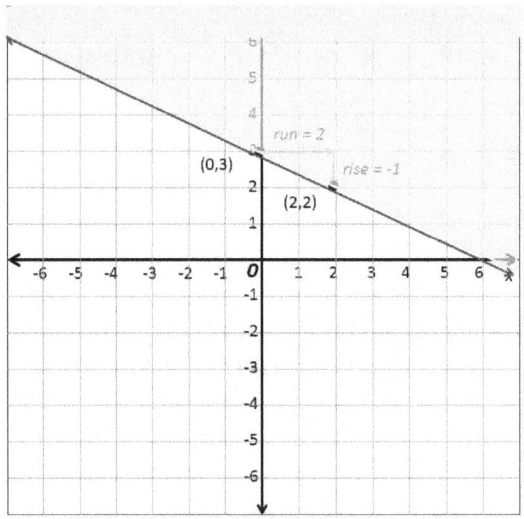

Figure A.1.4

The y-intercept (where x = 0) is the point (0, 3) will be the reference point. To graph the inequality, start at the reference point (0, 3) and go to the right (run) two (2) units and then go down (- rise) one (1) unit, since the slope is a $-\frac{1}{2}$ or the $\frac{rise}{run}$. A line passing through the two points (0, 3) and (2, 2) is the graph for the solution to the functional relationship and the area above the solid line must be shaded for the function $y \geq -\frac{1}{2}x + 3$.

KEEPING IT REAL - ALGEBRA I   A.1(C)

Example	Notes
$2x - 3y > 6$	Need to rewrite the inequality in the slope intercept form
$-2x + 2x - 3y > -2x + 6$	Additive Inverse; add $-2x$ to both sides
$-\dfrac{1}{3}(-3y > -2x + 6)$	Multiplicative Inverse, multiply by $-\dfrac{1}{3}$, to make coefficient of y one (1)
$y < \dfrac{2}{3}x - 2$	**Note: When multiplying an inequality by a negative quantity, you must change the > to < or in the other case, the < to >.**

Since the line is $y = \dfrac{2}{3}x - 2$ and the Slope Intercept form is $y < \dfrac{2}{3}x - 2$, the line cannot be included. Thus, the line must be dashed and shade the area below the dashed line, since the inequality is less than. The graph for the inequality $y < \dfrac{2}{3}x - 2$ is illustrated in Figure A.1.5 below. From the y-intercept (0, -2), the slope is $\dfrac{2}{3}$ or the $\dfrac{rise}{run}$.

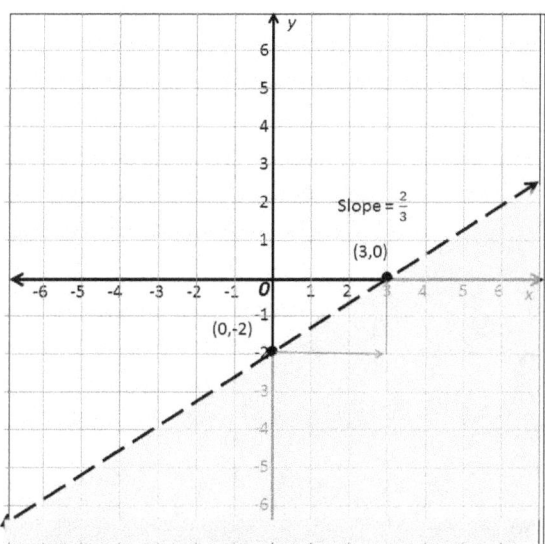

Figure A.1.5

KEEPING IT REAL - ALGEBRA I — A.1(C)

 INTEGRATING TECHNOLOGY

 CULTURAL COMPETENCY

 PROJECT-BASED LEARNING

 WRITING ACROSS THE CURRICULUM

www.keepingitrealeasymatch.com

KEEPING IT REAL - ALGEBRA I

A.1(C)

Student Materials

Objectives for A.1 (C) – Functional Relationships

1. Your family will travel 450 miles from home to reach Oklahoma. What inequality can be used to find all possible values of t, the time it will take your family to reach Oklahoma in hours, if you travel at an average speed of at least r miles per hour?

2. Write an equation for a function which the product of 5 and a number n is at least 50.

3. How would you express a number h, which is no less than 10 and no more than 45?

4. The top female basketball player score 400 points last season. If she plays 20 games this season, will an average of 25 points per game beat the record of last year?

5. Express the difference of a number t and 8, which is greater than 15 and less than 30.

KEEPING IT REAL - ALGEBRA I   A.1(C)

EasyMatch A.1 (D)

STAAR – Algebra I: Functional Relationships – the student will describe functional relationships in a variety of ways.

A.1 – Foundations for functions. The student understands that a function represents a dependence of one quantity on another and can be described in a variety of ways.

A.1 (D) – Represent relationships among quantities using [concrete] models, tables, graphs, diagrams, verbal descriptions, equations, and inequalities;
Readiness Standard

STANDARD AND LESSON OVERVIEW
TEACHER DIRECTIONS

A.1 (D) – Represent relationships among quantities using [concrete] models, tables, graphs, diagrams, verbal descriptions, equations, and inequalities;
Readiness Standard

Definition: A **relation** is any pairing of a set of inputs with a set of outputs. Every function is a relation, but not every relation is a function. A **relation is a function** if for every input there is exactly one output. For example, consider the two tables below>

a. **Table A – Input 3 has two different outputs, 2 and 0 – The relation is NOT a function.**

Input	1	2	3	3	5
Output	9	1	2	0	3

b. **Table B – Each Input has exactly one output – The relation is a function.**

Input	5	6	7	8	9
Output	4	3	2	1	0

A Vertical Line Test can be used to determine if a relation is a function. A relation represented by a graph is a function provided that no vertical line (up and down) passes through more than one point on the graph.

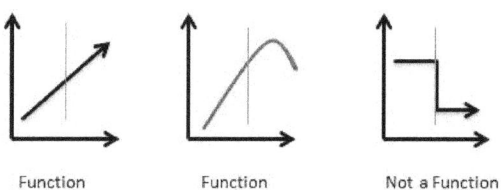

Function   Function   Not a Function

KEEPING IT REAL - ALGEBRA I   A.1(D)

 INTEGRATING TECHNOLOGY    CULTURAL COMPETENCY

 PROJECT-BASED LEARNING    WRITING ACROSS THE CURRICULUM

www.keepingitrealeasymatch.com

KEEPING IT REAL - ALGEBRA I   A.1(D)

Student Materials

Objectives for A.1 (D) – Relationships and Functions

1. Describe the relationship between x and y in the table below in equation form.

x	y
5	15
7	21
9	27
11	33

_____
_____
_____

2. Write an equation which represents the product of 5 and a number *n* added to 7 are equal to 57.

_____
_____
_____

3. Which data sets in the scatter plots below have a negative slope?

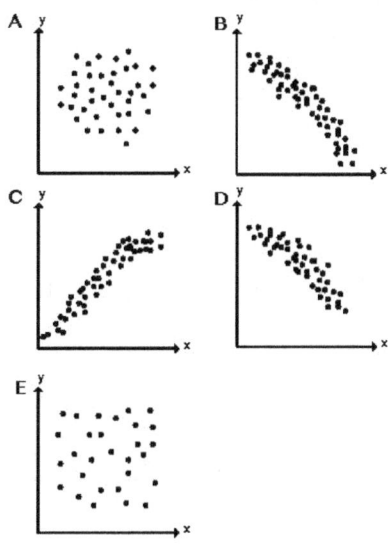

KEEPING IT REAL - ALGEBRA I

A.1(D)

4. Draw a graph of the function y > x + 2.

Insert Graph Below:

5. The graph of a quadratic function t passes through the points (-3, 31), (2, 3), (4, 13), and (6, 31). Use a table and a mapping diagram to represent the same relationship t.

Insert Graph Below:

6. Write an equation for the sum of a number r and 2 is greater than 5 and no more than 17.
_____
_____
_____

7. If the sum of 11 and a quantity 7 times a number j is equal to 58, what is the equation which represents the relationships?
_____
_____
_____

KEEPING IT REAL - ALGEBRA I   A.1(D)

8. Make a table for the function y = x − 3. The Domain: 11, 14, 15, 17

Insert Graph Below:

9. Graph the relationship y = 3x + 2.

Insert Graph Below:

10. Write an equation for the graph with the following points: (1, 2), (2, 3), (3, 4), (4, 5).
_____
_____
_____

> **EasyMatch A.1 (E)**
>
> **STAAR – Algebra I: Functional Relationships** – the student will describe functional relationships in a variety of ways.
>
> **A.1 – Foundation for functions.** The student understands that a function represents a dependence of one quantity on another and can be described in a variety of ways.
>
> **A.1 (E)** – Interpret and make decisions, predictions, and critical judgments from functional relationships. **Readiness Standard**

STANDARD AND LESSON OVERVIEW
TEACHER DIRECTIONS

A.1 (E) – Interpret and make decisions, predictions, and critical judgments from functional relationships. **Readiness Standard**

As illustrated in A.1 (D), functional relationships among quantities can be expressed using models, tables, graphs, diagrams, verbal descriptions, equations and inequalities. Often it is necessary to make interpretations, decisions, predictions and critical judgments from the relationships. When substituting a number for the variable in an open sentence like $x + 3 = 7$ or $3y > 7$, the resulting statement is either true or false. If the statement is true, the number is a **solution of the equation** or a **solution of the inequality**. There are other key vocabulary terms to introduce.

Definitions: The **best-fitting** line is the line that most closely follows a trend of the data. The process of finding the best-fitting line to model a set of data is called **linear regression**. Using a line or its equation to approximate a value between two known values is called **linear interpolation**. Using a line or its equation to approximate a value outside the range of known values is called **linear extrapolation**. Recall that a **scatter plot** is a type of display for paired data, where each data pair is plotted as a point on a graph.

Interpolate Using an Equation

Situation: The table below shows the total number of T-shirts ordered for the senior class in hundreds for recent years. Approximate the number of T-Shirts ordered in 2008.

Year	2005	2007	2009	2011	2013
T-Shirts (100)	2	3	5	6	9

KEEPING IT REAL - ALGEBRA I   A.1(E)

Solution process:

    a. Make a scatter plot

    b. Find an equation that models the T-shirts ordered

    c. Graph the best-fitting line and approximate the value.

Extrapolate Using an Equation

Situation: Use the equation from the T-Shirt table above to approximate the number of T- Shirts to order in 2014 and 2015.

    a. Evaluate the equation of the best-fitting line for 2014 and 2015.

    b. Use the graph of the best-fitting line to predict the number of T-Shirts as well.

 INTEGRATING TECHNOLOGY    CULTURAL COMPETENCY

 PROJECT-BASED LEARNING    WRITING ACROSS THE CURRICULUM

www.keepingitrealeasymatch.com

KEEPING IT REAL - ALGEBRA I — A.1(E)

Student Materials

Objectives for A.1 (E) — Relationships and Functions

1. The enrollment of high school students is currently 900. The function $s = 900 + 4h^2$ can be used to estimate the enrollment of students in the high school h years from now. Based on this function, in how many years will the student enrollment reach 1000 students?

2. Your dishwasher at home must be loaded with the same number of dishes each time it is used to conserve water and energy. The table below shows the total number of dishes washed as a function of the number of times used. Based on the data, what is the total number of dishes washed in 4 uses?

Number of Times Used	Total Number of Dishes Washed
2	36
3	39

3. The table below shows the number of participants in the city little league baseball program for the period 2008 – 2012. Predict the year in which the number of little league baseball participants will reach 500.

Year	2008	2009	2010	2011	2012
Participants	200	250	300	350	400

KEEPING IT REAL - ALGEBRA I   A.1(E)

4. Make a scatter plot of the data and find the equation for the best-fitting line. Approximate the value of y for x = 5.

x	0	2	4	6	7
y	2	7	14	17	20

Insert the scatter plot below.

5. Make a scatter plot of the data and find the equation for the best-fitting line. Approximate the value of y for x = 10.

x	1	3	5	7	9
y	0.4	1.4	1.9	2.3	3.2

Reporting Category 2

EasyMatch A.2 (A)

**STAAR – Algebra I: Properties and Attributes of Functions -
The student will demonstrate an understanding of the properties and attributes of functions.**

A.2 -Foundations for functions. The student uses the properties and attributes of functions. The student is expected to

A.2 (A) - Identify and sketch the general forms of linear (y = x) and quadratic (y = x²) parent functions; **Supporting Standard**

STANDARD AND LESSON OVERVIEW
TEACHER DIRECTIONS

A.2 (A) - Identify and sketch the general forms of linear (y = x) and quadratic (y= x²) parent functions; **Supporting Standard**

A **family of functions** is a group of functions with similar characteristics. For example, functions that have the form $f(x) = mx + b$ constitute a family of linear functions. The power of y and x are both raised to the first power.

The most basic linear function in the family of all linear functions, called the **parent linear function** is: $f(x) = x$, where $f(x)$ represents the output **y**.

The Identity Function (y = x) is an increasing function and the **slope** (ratio of the vertical rise over the horizontal run $-\frac{rise}{run}$) is positive. The statement that **f** is an increasing function means that if x_1 and x_2 are in the Domain of **f** and x_2 is greater than x_1, then $f(x_2)$ is greater than $f(x_1)$. Thus, the **Identity Function** is an **Increasing Function**. When x_2 is greater than x_1 then the height of $f(x_2)$ is greater than the height of $f(x_1)$. From left to right, the function is increasing (going up) – note **Figure A.2.1A**.

KEEPING IT REAL - ALGEBRA I A.2(A)

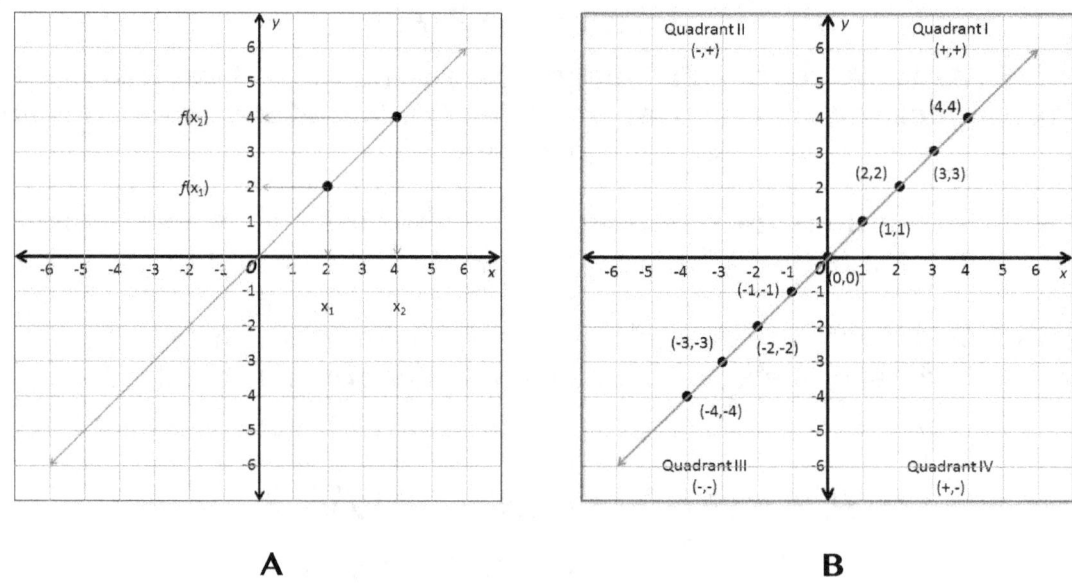

Figure A.1.1 - The Identity Function

Another approach supporting that the function y = x is the Identity Function is because for each ordered pair, x and y are the same. Another plot of the Identity Function is illustrated in Figure A.2.1B. Note that for each ordered pair, the values for x and y are the same.

Quadratic Functions and Equations

A quadratic function is a nonlinear function that can be written in the standard form $y = ax^2 + bx + c$, where $a \neq 0$. Every quadratic function has a U-shaped graph called a parabola. Given the parent quadratic function, $y = x^2$, if the coefficient of the x^2 term (a) is positive the graph will turn up at the vertex, the lowest or highest point of the parabola. The vertex for the graph of $y = x^2$ is (0, 0).

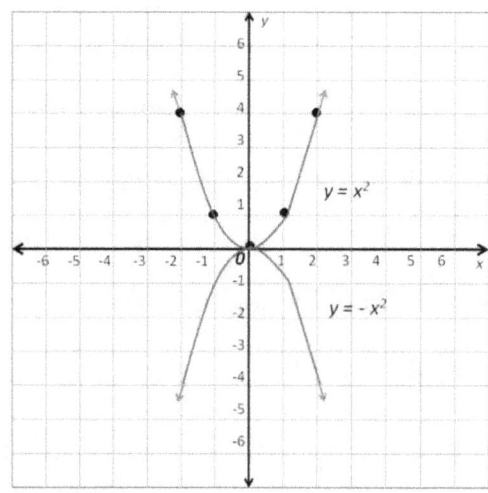

Figure A.2.2 $y = x^2$ and $y = -x^2$

KEEPING IT REAL - ALGEBRA I A.2(A)

The domain of the $y = x^2$ function is all Real Numbers. The range is all $y \geq 0$. If $y = -x^2$, the graph turns down and the domain is all Real Numbers and the range is all $y \leq 0$.

 INTEGRATING TECHNOLOGY

 CULTURAL COMPETENCY

 PROJECT-BASED LEARNING

 WRITING ACROSS THE CURRICULUM

www.keepingitrealeasymatch.com

KEEPING IT REAL - ALGEBRA I   A.2(A)

Student Materials

Objectives for A.2 (A) — Linear and Quadratic Parent Functions

What is the parent function for a graph of the set of ordered pairs below:
$\{(3, 11), (-1, 3), (-7, -9), (-4, -3)\}$
Insert your answer below:

2. Graph the equation $y = x^2$ with the domain: 1, -1, 2, -2, 3, -3.

KEEPING IT REAL - ALGEBRA I A.2(A)

3. Graph the equation $y = 2x^2$ for the same domain in problem 2.

4. Graph the equation $y = 3x^2$ for the same domain in problem 2.

5. What observation is made when comparing the graphs of problems 2, 3 and 4?

_____
_____
_____
_____

KEEPING IT REAL - ALGEBRA I

A.2(A)

> ### EasyMatch A.2 (B)
>
> **STAAR – Algebra I: Properties and Attributes of Functions – The student will demonstrate an understanding of the properties and attributes of functions.**
>
> **A.2 -Foundations for functions.** The student uses the properties and attributes of functions. The student is expected to
>
> A.2 (B) – Identify mathematical domains and ranges and determine reasonable domain and range values for given situations, both continuous and discrete;
> <u>**Readiness Standard**</u>

STANDARD AND LESSON OVERVIEW
TEACHER DIRECTIONS

A.2 (B) – Identify mathematical domains and ranges and determine reasonable domain and range values for given situations, both continuous and discrete;
<u>**Readiness Standard**</u>

Mathematical Domains and Ranges

The **Domain** of the function is the called the **Independent Variable (x)** and the **Range** of the function is called the **Dependent Variable (y).** The values for the function are identified **(domain – inputs – x-values)** and then determine the results **(range – outputs – y-values)** for the function. For the function $y = x$; if $x = 1$, then $y = 1$; if $x = 2$, then $y = 2$; if $x = 3$, then $y = 3$; and if $x = 4$, then $y = 4$. Thus, the function $y = x$ consists of the order pairs (0, 0), (1, 1), (2, 2), (3, 3), (4, 4) ... and also extends in the direction of (-1, -1), (-2, -2), (-3, -3) and (-4, -4).... . For $y = x$, the **Domain (x-values)** is all Real Numbers and the **Range (y-values)** is all Real Numbers.

Continuous and Discrete Situations

The graph of a function can consist of individual points or the graph of a function can also be a line or part of a line with no breaks. A **discrete function** has a graph that consists of isolated points. A **continuous function** has a graph that is not broken, see Figure A.2.3.

KEEPING IT REAL - ALGEBRA I   A.2(B)

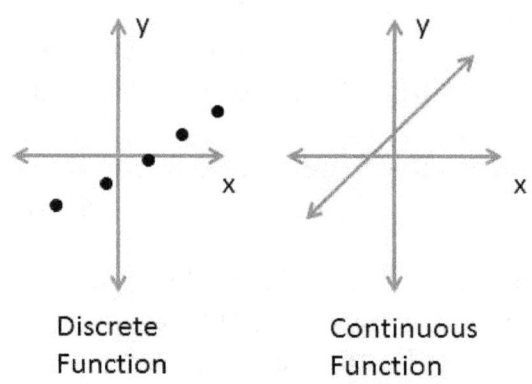

Figure A.2.3 - Discrete - Continuous Functions

As a general rule, you can determine that a function is continuous, if you do not have to lift your pencil from the paper to draw its graph.

KEEPING IT REAL - ALGEBRA I   A.2(B)

Student Materials

Objectives for A.2 (B) – Continuous and Discrete Functions

1. Draw a graph which shows a function with a domain of all real numbers greater than 7.

2. Identify the range of the function shown below.

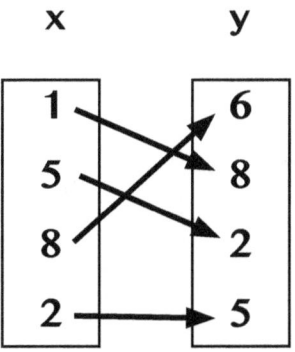

___
___
___
___

KEEPING IT REAL - ALGEBRA I   A.2(B)

3. Determine whether the function represented by the table below is discrete or continuous. If the function is continuous, graph the function and find the value of y when x = 2.5 and when x = 3.5.

x	1	2	3
y	0.5	1.0	1.5

Graph the function with the given domains. Classify the function as discrete or continuous.

a. $y = -3x + 2$; Domain: -2, -1, 0, 1, 2

b. $y = 3x - 2$; Domain: $x \leq 0$.

c. $y = 0.5x$; Domain: -2, -1, 0, 1, 2

d. $y = \frac{1}{3}x + \frac{2}{3}$; Domain: $x \geq -2$

e. $y = x$; Domain: $x > 0$

> ### EasyMatch A.2 (C)
>
> **STAAR – Algebra I: Properties and Attributes of Functions** - The student will demonstrate an understanding of the properties and attributes of functions.
>
> **A.2 -Foundations for functions.** The student uses the properties and attributes of functions. The student is expected to
>
> A.2 (C) – Interpret situations in terms of given graphs or create situations that fit given graphs; **Supporting Standard**

STANDARD AND LESSON OVERVIEW
TEACHER DIRECTIONS

A.2 (C) – Interpret situations in terms of given graphs or create situations that fit given graphs; **Supporting Standard**

Graphs are used to represent various functions. Functions can be represented in tables, sets of ordered pairs or paired data, an equation or word descriptions, just to name a few. The graph is often labeled with the x-axis (horizontal) and the y-axis (vertical). Both the x and y axes have arrows at each end which indicates that the x-axis is infinite in both the positive (right) and negative (left) horizontal direction and the y-axis is infinite in both the positive (up) and negative (down) directions.

A **scatter plot** is a type of display for paired data, where each data pair is plotted as a point on a graph – used to represent sets of data. Scatter plots are also used to show trends in data, relationship between paired data or correlation between the data. When data show a positive or negative correlation, model the trend in the data using a **line of fit**. A **line of fit** is drawn in a way that it appears to fit the data closely. There should be approximately the same number of points above the line as below the line.

KEEPING IT REAL - ALGEBRA I — A.2(C)

For example:

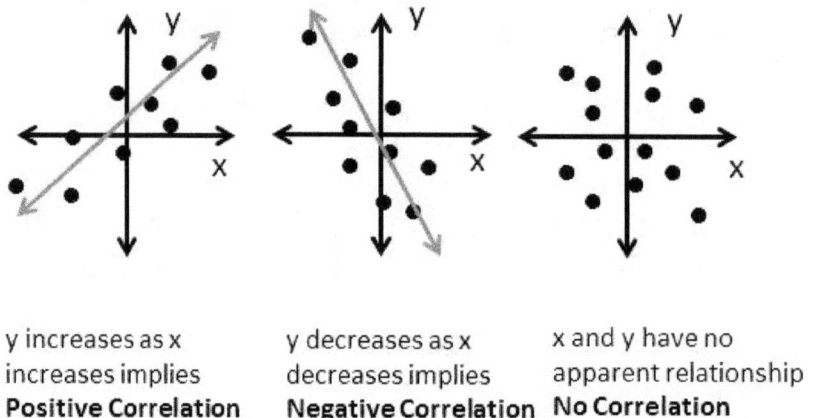

Figure A.2.4 - Scatter Plots

Functions can also be discrete or continuous, with a positive, negative or zero slope

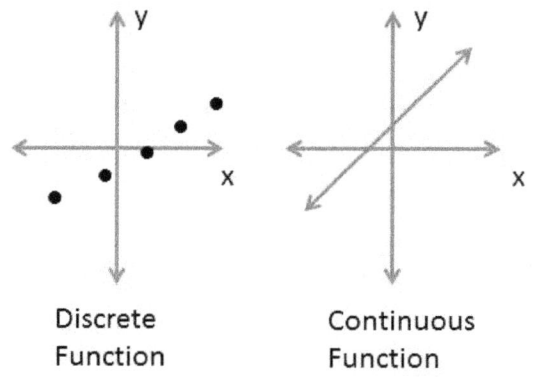

Figure A.2.5 - Discrete - Continuous Functions

www.keepingitrealeasymatch.com

KEEPING IT REAL - ALGEBRA I

A.2(C)

Student Materials

Objectives for A.2 (C) – Interpreting Situations Using Graphs

1. Interpret the situations for the graphs below:

a. Discuss A, B, C, D and E

b.

c.

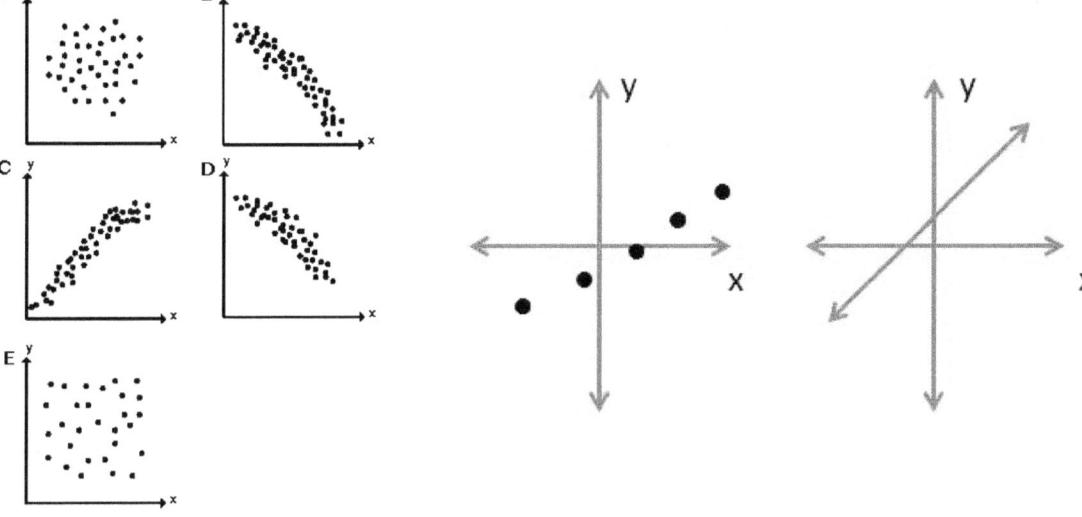

d.

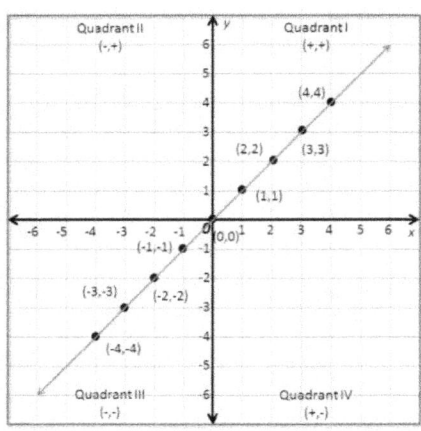

e.

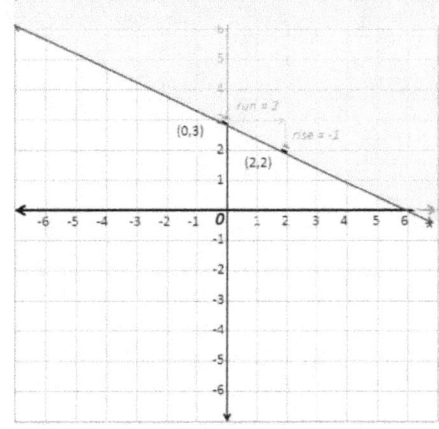

KEEPING IT REAL - ALGEBRA I

A.2(C)

EasyMatch A.2 (D)

STAAR – Algebra I: Properties and Attributes of Functions - The student will demonstrate an understanding of the properties and attributes of functions.

A.2 -Foundations for functions. The student uses the properties and attributes of functions. The student is expected to

A.2 (D) – Collect and organize data, make and interpret scatterplots (including recognizing positive, negative, or no correlation for data approximating linear situations), and model, predict, and make decisions and critical judgments in problem situations. **Readiness Standard**

STANDARD AND LESSON OVERVIEW
TEACHER DIRECTIONS

A.2 (D) – Collect and organize data, make and interpret scatterplots (including recognizing positive, negative, or no correlation for data approximating linear situations), and model, predict, and make decisions and critical judgments in problem situations. **Readiness Standard**

A Scatter Plot is a type of display for paired data, where each data pair is plotted as a point on a graph – used to represent sets of data. Scatter plots are also used to show trends in data, relationship between paired data or correlation between the data. When data show a positive or negative correlation, model the trend in the data using a line of fit. A line of fit is drawn in a way that it appears to fit the data closely. There should be approximately the same number of points above the line as below the line. The line of fit can be used to write an equation which describes the relationship between the paired data.

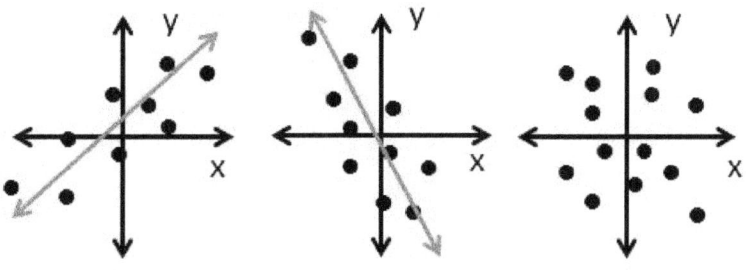

y increases as x increases implies **Positive Correlation**

y decreases as x decreases implies **Negative Correlation**

x and y have no apparent relationship **No Correlation**

Figure A.2.6 - Scatter Plots

KEEPING IT REAL - ALGEBRA I A.2(D)

Note: The line below fits the data which represents the graph of $y = x$.

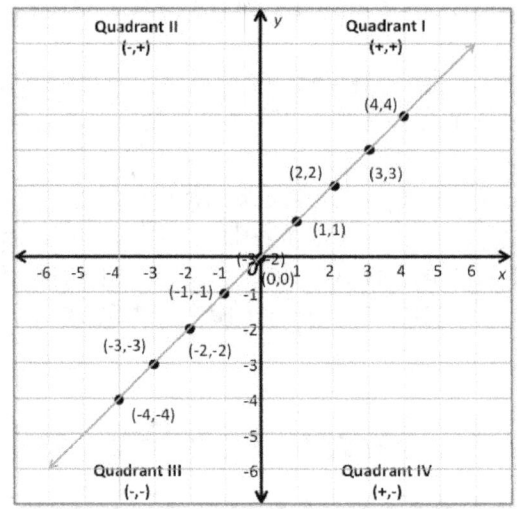

 INTEGRATING TECHNOLOGY

 CULTURAL COMPETENCY

 PROJECT-BASED LEARNING

 WRITING ACROSS THE CURRICULUM

www.keepingitrealeasymatch.com

KEEPING IT REAL - ALGEBRA I

A.2(D)

Student Materials

Objectives for A.2 (D) – Interpreting Scatter Plots and Situations Using Graphs

Interpret the scatterplots below. Identify the positive and negative slopes or no correlation. Predict, make decisions or judgments as appropriate.

a.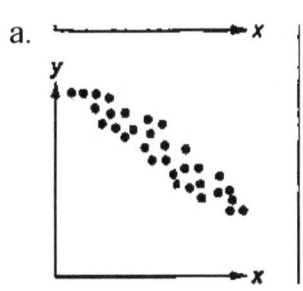

KEEPING IT REAL - ALGEBRA I   A.2(D)

b.

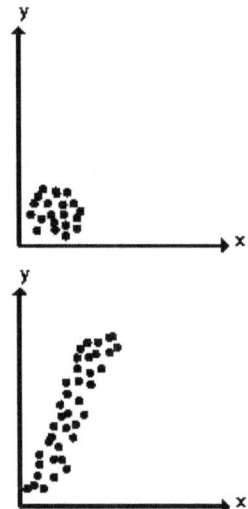

> ### EasyMatch A.3 (A)
>
> **STAAR – Algebra I: Properties and Attributes of Functions - The student will demonstrate an understanding of the properties and attributes of functions.**
>
> **A.3 - Foundations for functions.** The student understands how algebra can be used to express generalizations and recognizes and uses the power of symbols to represent situations. The student is expected to
>
> A.3 (A) - Use symbols to represent unknowns and variables;
> **Supporting Standard**

STANDARD AND LESSON OVERVIEW
TEACHER DIRECTIONS

A.3 (A) - Use symbols to represent unknowns and variables; **Supporting Standard**

A **variable** is a letter used to represent one or more numbers. The numbers are the values for the variables. **Algebraic expressions** or **variable expressions** consist of numbers, variables, and operations.

When evaluating an algebraic expression, substitute a number for each variable, perform the operation(s), and simplify the result, if necessary.

In solving equations or evaluating algebraic expressions, there are three very important mathematical properties to introduce which will provide consistency and structure to solving functional relationships expressed as **equations** or **inequalities**. The three properties are the **Additive Inverse, Multiplicative Inverse** and **Distributive Property**.

The **Additive Inverse** is introduced as a one-step equation involving addition. Rational numbers will be included to reinforce concepts introduced in prior mathematics courses.

Definition: Additive Inverse property states that if a is a real number then there exist a negative a (-a) such that a + -a = 0. **The opposite of *a* is its additive inverse (-a)**, such that the sum of a number a and its opposite is 0.

KEEPING IT REAL - ALGEBRA I — A.3(A)

Example   **Notes**

$x - \dfrac{2}{3} = \dfrac{-3}{5}$ The objective is to get x by itself on the left side of the = sign

$x - \dfrac{2}{3} + \dfrac{2}{3} = \dfrac{-3}{5} + \dfrac{2}{3}$ The additive inverse of $-\dfrac{2}{3}$ is $+\dfrac{2}{3}$, added to both sides

$x + 0 = \dfrac{-9 + 10}{15}$ Recall, $\dfrac{a}{b} + \dfrac{c}{d} = \dfrac{ad + cb}{bd}$, bd is a common denominator

$x = \dfrac{1}{15}$ 0 is the identity element of addition thus, $x + 0 = x$

Multiplicative Inverse

The **Multiplicative Inverse** involves another one step equation involving multiplication. Again, rational numbers will be included to reinforce previous concepts.

Definitions: The **Coefficient** of x, is the number to the left of x. Consider a value expressed by 2x (2 times x), where 2 is the coefficient of x. It can also be expressed that the coefficient is the number in front of x. The **Multiplicative Inverse** property states that if a is a real number and a is not zero (0), then there exists another number, $\dfrac{1}{a}$, the reciprocal of a, such that $a \times \dfrac{1}{a}$ or $\dfrac{1}{a}(a) = \dfrac{a}{a} = 1$. Zero (0) does not have a multiplicative inverse because there is no number a, such that $0 \times a = 1$. Since one (1) is the identity element for multiplication, the multiplicative inverse is a very powerful property of mathematics used in solving various types of equations.

Example   **Notes**

$\dfrac{2}{3}x = \dfrac{1}{4}$ Solving for x, the objective is to make the coefficient of x one (1)

$\dfrac{3}{2} \times \dfrac{2}{3}x = \dfrac{1}{4} \times \dfrac{3}{2}$ Multiply both sides by the Multiplicative Inverse of $\dfrac{2}{3}$, which is $\dfrac{3}{2}$

$\dfrac{6}{6}x = \dfrac{3}{8}$ Based on properties of rational numbers $\dfrac{a}{b} \times \dfrac{c}{d} = \dfrac{ac}{bd}, \dfrac{6}{6} = 1$

$x = \dfrac{3}{8}$ Since 1 is the Identify element of multiplication, solved for x

KEEPING IT REAL - ALGEBRA I A.3(A)

Distributive Property

Definitions: The **Distributive Property** of Real Numbers states that if *a*, *b* and *c* are Real Numbers, then $a(b+c) = a \times b + a \times c$. The number *a* is multiplied or distributed to both *b* and *c*. That is,

$$2(3+4) = 2 \times 3 + 2 \times 4$$
$$2(7) = 6 + 8$$
$$14 = 14$$

 INTEGRATING TECHNOLOGY

 CULTURAL COMPETENCY

 PROJECT-BASED LEARNING

 WRITING ACROSS THE CURRICULUM

www.keepingitrealeasymatch.com

KEEPING IT REAL - ALGEBRA I   A.3(A)

Student Materials

Objectives for A.3 (A) – Symbols and Variables

1. You are the manager at the local McDonalds and you begin each shift with the same total amount of money. You keep $500 in a safe and distribute the rest equally to the 3 cashiers in the restaurant. Your situation can be represented by the function $y = \dfrac{x - 500}{3}$. What does the variable x represent in your situation? What does the variable y, represent in your situation?

___

2. The total cost of seeing a play at a theater can be represented by the expression $a + r$, where a is the cost (in dollars) of admissions and r is the cost (in dollars) for refreshments and tips. Suppose you pay $15.00 for admission and $12.50 for refreshments, what is your total cost for the play?

___

3. Suppose you decide to take a date to the play, what would the total cost be?

___

KEEPING IT REAL - ALGEBRA I   A.3(A)

4. Evaluate the expressions below when $s = 5$.

    a. $13s$

    b. $\dfrac{15}{s}$

    c. $s - 1$

    d. $23 - s$

    e. $4s + 2s - s$

    f. $s(4+3) - 10$

5. Solve each equation for the given variable.

    a. $4(x - 6) = 18$

    b. $15a + 3 = -22$

    c. $b + 9 - 2b = 3$

    d. $2k + 3(k+5) = -5$

    e. $5(m - 2) + 3m = 22$

    f. $7t - 4 + 2t = -15$

    g. $11k - 3(k + 2) = 55$

KEEPING IT REAL - ALGEBRA I

A.3(A)

EasyMatch A.3 (B)

STAAR – Algebra I: Properties and Attributes of Functions - The student will demonstrate an understanding of the properties and attributes of functions.

A.3 - Foundations for functions. The student understands how algebra can be used to express generalizations and recognizes and uses the power of symbols to represent situations. The student is expected to

A.3 (B) – Look for patterns and represent generalizations algebraically.
Supporting Standard

STANDARD AND LESSON OVERVIEW
TEACHER DIRECTIONS

A.3 (B) – Look for patterns and represent generalizations algebraically. **Supporting Standard**

Patterns involve a form or model proposed to imitate or replicate. Patterns are also used to represent a group of data points or a series of operations. Scatter plots are another form of patterns used to represent generalizations algebraically.

Finding the pattern in numbers is a skill that lays the foundation for data analysis. The numbers in series can range from simple addition or subtraction patterns (at the easy level) to rolling mixed computations (at the complex level).

 INTEGRATING TECHNOLOGY    CULTURAL COMPETENCY

 PROJECT-BASED LEARNING    WRITING ACROSS THE CURRICULUM

www.keepingitrealeasymatch.com

KEEPING IT REAL - ALGEBRA I   A.3(B)

Student Materials

Objectives for A.3 (B) – Patterns

1. If the first six numbers in a pattern are: $\frac{1}{3}, \frac{4}{3}, 3, \frac{16}{3}, \frac{25}{3}, 12, \ldots$ What expression can be used to find the nth number?

_____

_____

_____

2. Complete the number sequences with the values that should come next.

   a.   4, 7, 10, 13, 16, 19, 22, 25, 28, _____, _____, _____

   b.   6, 4, 8, 6, 10, 8, 12, 10, 14, _____, _____, _____

   c.   7, 8, 9, 10, 11, 12, 13, 14, 15, _____, _____, _____

   d.   9, 9, 10, 11, 11, 12, 13, 13, 14, _____, _____, _____

   e.   1, 2, 3, 5, 8, 13, 21, 34, 55, _____, _____, _____

   f.   1, 3, 4, 7, 11, 18, 29, 47, 76, _____, _____, _____

   g.   14, 13, 12, 11, 10, 9, 8, 7, 6, _____, _____, _____

   h.   7, 7, 5, 3, 3, 1, -1, -1, -3, _____, _____, _____

   i.   7, 7, 5, 3, 3, 1, -1, -1, -3, _____, _____, _____

   j.   30, 62, 94, 126, 158, 190, 222, 254, 286, _____, _____, _____

EasyMatch A.4 (A)

STAAR – Algebra I: Properties and Attributes of Functions - The student will demonstrate an understanding of the properties and attributes of functions.

A.4 – Foundations for functions. The student understands the importance of the skills required to manipulate symbols in order to solve problems and uses the necessary algebraic skills required to simplify algebraic expressions and solve equations and inequalities in problem situations. The student is expected to

A.4 (A) – Find specific function values, simplify polynomial expressions, transform and solve equations, and factor as necessary in problem situations; **Readiness Standard**

STANDARD AND LESSON OVERVIEW
TEACHER DIRECTIONS

A.4 (A) – Find specific function values, simplify polynomial expressions, transform and solve equations, and factor as necessary in problem situations; **Readiness Standard**

Exponents, Monomials and Polynomials

Definitions: A power is an expression that represents repeated multiplication of the same factor. For example, 81 is a power of 3 because $81 = 3 \times 3 \times 3 \times 3$. A power can be written in a form using two numbers, a base and an exponent. The exponent represents the number of times the base is used as a factor, so 81 can be written as 3^4, where 3 is the base and 4 is the exponent. X^5 means that x is the base and 5 is the exponent.

Expanding $x^5 = x \times x \times x \times x \times x$.

For $x^5 \times x^3 = x \times x \times x \times x \times x \times x \times x \times x$. There are 8 x's which implies that $x^5 \times x^3 = x^8$. So, in multiplying exponents with the same base, the exponents are added such that $x^n \times x^m = x^{n+m}$.

In dividing with the same base, $\dfrac{x^5}{x^3} = \dfrac{xxxxx}{xxx} = x^{5-3} = x^2$, therefore $\dfrac{x^n}{x^m} = x^{n-m}$.

For $(x^2)^3$, implies that x^2 is multiplied by itself 3 times illustrated as $x^2 \times x^2 \times x^2 = x^{2+2+2} = x^6$

Therefore $(x^n)^m = x^{n \times m}$.

KEEPING IT REAL - ALGEBRA I   A.4(A)

Note: Any number divided by itself, other than zero (0), is equal to one (1). Since $1 = \dfrac{x^n}{x^n}$, which can be written as $x^{n-n} = x^0$, therefore any x^0 power is equal to one (1).

Given a number x^n and multiplying x^n by one (1) expressed as $\dfrac{x^n}{x^n}$, hence $x^{-n} = x^{-n} \times \dfrac{x^n}{x^n} = \dfrac{x^0}{x^n}$.

Since $x^0 = 1$, then $x^{-n} = \dfrac{1}{x^n}$.

Given $\dfrac{1}{x^{-n}}$ and multiplying $\dfrac{1}{x^{-n}}$ by one in the form of $\dfrac{x^n}{x^n}$ gives, $\dfrac{1}{x^{-n}} = \dfrac{1}{x^{-n}} \times \dfrac{x^n}{x^n} = \dfrac{x^n}{x^0}$.

Since any number to the zero (0) power is one, then $\dfrac{1}{x^{-n}} = \dfrac{x^n}{x^0} = \dfrac{x^n}{1} = x^n$.

Therefore $\dfrac{1}{x^{-n}} = x^n$.

Given $\dfrac{1}{x^n}$ multiplied by one (1) expressed as $\dfrac{x^{-n}}{x^{-n}}$ then $\dfrac{1}{x^n} = \dfrac{1}{x^n} \times \dfrac{x^{-n}}{x^{-n}} = \dfrac{x^{-n}}{x^0} = \dfrac{x^{-n}}{1} = x^{-n}$.

Therefore, $\dfrac{1}{x^n} = x^{-n}$.

If given $\sqrt[n]{x^m} = x^{m/n}$, take the n root of the number and raise that number to the m power.

If there is no *n* in the index, then the square root is implied.

Example: $\sqrt[3]{8^4} = x^{4/3}$ which implies to take the cube root of 8 and raise it to the fourth power.

Must Know the Rules and Principles of Exponents:

$$x^n \times x^m = x^{n+m} \qquad \dfrac{x^n}{x^m} = x^{n-m}$$

$$(x^n)^m = x^{n \times m} \qquad x^0 = 1$$

$$x^{-n} = \dfrac{1}{x^n} \qquad x^n = \dfrac{1}{x^{-n}}$$

$$\sqrt[n]{x^m} = x^{m/n}$$

A **monomial** is a number, a variable, or the product of a number and one or more variables with a whole number exponent. The **degree of the monomial** is the sum of the exponents of the variables in the monomial. The degree of a nonzero constant term is 0. The constant 0 does not have a degree, e. g. 10, 3x,

KEEPING IT REAL - ALGEBRA I   A.4(A)

Examples

Monomial	Degree
9	0
2x	1
$\frac{1}{3}ab^2$	1+2=3
$-1.6m^5$	5

Not a Monomial	Reason
4+x	A sum is not a monomial
$\frac{1}{n}$	1 monomial cannot have a variable in the denominator
5^a	A monomial cannot have a variable exponent
x^{-1}	The variable must have a whole number exponent.

A **polynomial** is a monomial or a sum of monomials, each called a term, of the polynomial. The **degree of a polynomial** is the greatest degree of its terms e. g. $2x^3 + x^2 - 5x + 12$, the greatest degree is 3, so the degree of the polynomial is 3. The leading coefficient is 2.

Examples

Expression	Is it a polynomial?	Classify by degree and number of terms
8	yes	0 degree monomial
$2x^2 + x - 4$	yes	2nd degree trinomial
$5n^4 - 7^n$	No; variable exponent	
$n^{-2} - 2$	No; negative exponent	
$4bc^3 + 4b^4c$	Yes	4th degree binomial

KEEPING IT REAL - ALGEBRA I   A.4(A)

In performing operations on polynomials, the following will be helpful:

Addition: To add polynomials, add like terms.

Subtraction: To subtract a polynomial, add its opposite. To find the opposite of a polynomial, multiply each of its terms by -1.

Multiplying Polynomials: Several methods are available for multiplying polynomials, with each method based on the **Distributive Property**.

Factoring: A key concept in factoring is the **Zero-Product Property**, which by definition: Let a and b be real numbers. If ab = 0, then a = 0 or b = 0. The **zero-product property** is used to solve an equation when one side is zero (0) and the other side is a product of polynomial factors. In **factoring** – write the equation as a product of other polynomials. Look for the greatest common factor (GCF) of the polynomial's terms. This is a monomial with and integer coefficient that divides evenly into each term. The solutions of such an equation are also called **roots**.

For the quadratic equation:

Notes

$x^2 - 2x - 3 = 0$	**Factoring** – write as a product of other polynomials
$(x - 3)(x + 1) = 0$	Zero-product property
$(x-3) = 0$ or $(x+1) = 0$	Solved for x: The solutions of the equation are 3 and -1
$x = 3$ or $x = -1$	Thus, the roots are 3 and -1.

Examples

1. $(2x^3 - 5x^2 + x) + (2x^2 + x^3 - 1) = 3x^3 - 3x^2 + x - 1$

2. $(4n^2 + 5) - (-2n^2 + 2n - 4) = (4n^2 + 5) + (2n^2 - 2n + 4) = 6n^2 - 2n + 9$

3. $2x^3(x^3 + 3x^2 - 2x + 5) = 2x^3(x^3) + 2x^3(3x^2) - 2x^3(2x) + 2x^3(5) = 2x^6 + 6x^5 - 4x^4 + 10x^3$

4. $(x - 3)(2x + 3)$ or $[x + (-3)](2x + 3) = 2x^2 + 3x - 6x - 9 = 2x^2 - 6x - 9$

5. $(2b^2 + 3b - 4)(2b - 3) =$

$$\begin{array}{r} 2b^2 + 3b - 4 \\ \times \quad 2b - 3 \\ \hline -6b^2 - 9b + 12 \\ 4b^3 + 6b^2 - 8b \quad\quad \\ \hline 4b^3 \quad\quad\; -17b + 12 \end{array}$$

KEEPING IT REAL - ALGEBRA I   A.4(A)

 INTEGRATING TECHNOLOGY    CULTURAL COMPETENCY

 PROJECT-BASED LEARNING    WRITING ACROSS THE CURRICULUM

www.keepingitrealeasymatch.com

KEEPING IT REAL - ALGEBRA I

A.4(A)

Student Materials

Objectives for A.4 (A) – Function Values and Polynomials Expressions

1. The perimeter of one side of a box is 42 inches. The length of the box can be represented by (x+4), and its width can be represented by (2x – 7). What are the length and width of the side of the box in inches?

2. Find an equivalent inequality for $7x - 2y \geq 8$.   (Use y = mx + b form)

3. Find an equivalent expression for $-6x^2 - 11x - 4$ by factoring.

4. Solve the following polynomial equations:

   a. $5(x+2) - 4(x-1) = 24$

   b. $2(y-5) - (y+6) = 4$

KEEPING IT REAL - ALGEBRA I   A.4(A)

c. $2(3n-1) - (n+6) = 7$

d. $x^2 - 4x - 5 = 0$

e. $2(y+1) + 3(y-1) = 9$

f. $x^2 + 5x + 6$

g. $x^2 + 11x + 18$

h. $x^2 + 3x + 2$

KEEPING IT REAL - ALGEBRA I   A.4(A)

> ### EasyMatch A.4 (B)
>
> **STAAR – Algebra I: Properties and Attributes of Functions** - The student will demonstrate an understanding of the properties and attributes of functions.
>
> **A.4 – Foundations for functions.** The student understands the importance of the skills required to manipulate symbols in order to solve problems and uses the necessary algebraic skills required to simplify algebraic expressions and solve equations and inequalities in problem situations. The student is expected to
>
> A.4 (B) – Use the commutative, associative, and distributive properties to simplify algebraic expressions; **Supporting Standard**

STANDARD AND LESSON OVERVIEW
TEACHER DIRECTIONS

A.4 (B) – Use the commutative, associative, and distributive properties to simplify algebraic expressions; **Supporting Standard**

Algebraic expressions are evaluated using key properties. The commutative, associative and distributive properties are of interest in this content expectation. The properties are very similar for both addition and multiplication:

Commutative Property:
The order of adding two numbers does not change the sum: a+b = b+a.
The order of multiplying two numbers does not change the product: a x b = b x a

Associative Property:
The way three numbers are grouped in a sum does not change the sum: (a+b)+c = a+(b+c)
The way three numbers are grouped in a product does not change the product:
(a x b) x c = a x (b x c)

Distributive Property: Let a, b and c be real numbers:
The product of a and (b+c) : a(b+c) = ab + ac or (b +c)a = ba + ca
The product of a and (b-c): a(b-c) = ab – ac or (b-c)a = ba – ca

KEEPING IT REAL - ALGEBRA I   A.4(B)

Examples

Commutative Property: $3 + 2 = 2 + 3$
                                       $3 \times 3 = 2 \times 3$

Associative Property: $(3 + 2) + 1 = 3 + (2 + 1)$
                               $(3 \times 2) \times 1 = 3 \times (2 \times 1)$

Distributive Property: $2(3 + 4) = 2 \times 3 + 2 \times 4$ $2(4 - 3) = 2 \times 4 - 2 \times 3$
                              $(3 + 4) \times 2 = 3 \times 2 + 4 \times 2$ $(4 - 3)2 = 4 \times 2 - 3 \times 2$

When simplifying algebraic expression, ensure the following **Order of Operations** is followed:

     1. **P**arentheses
     2. **E**xponents
     3. **M**ultiplication and **D**ivision
     4. **A**ddition and **S**ubtraction
     (PEMDAS)

     5. Multiplication and Division (in order from left to right when both operations are in the same problem.)
     6. Addition and Subtraction (in order from left to right when both operations are in the same problem.)

 INTEGRATING TECHNOLOGY    CULTURAL COMPETENCY

 PROJECT-BASED LEARNING    WRITING ACROSS THE CURRICULUM

www.keepingitrealeasymatch.com

KEEPING IT REAL - ALGEBRA I   A.4(B)

Student Materials

Objectives for A.4 (B) – Simplify Algebraic Expressions

1. $1.5(-6d + 4) - 2(d + 3) + 5(d - 2) =$

2. $8 - 4 \times 3(4 - 3) + 2 =$

3. $3(x + 2) = 2x + 9$

4. $7(6 \div 2) + 3(-2 \times 2) =$

5. $5(3 \times 2) - 10 \times 0 =$

6. $3^2 + 4^2 - 2^3 =$

KEEPING IT REAL - ALGEBRA I   A.4(B)

7. $\dfrac{(9-5) \times 4}{(6-4) \times 2} + 3 =$

8. $2(0-2) + 4 \div 2 =$

9. $40 \div (2+3) \times 2 =$

10. $8 - (2 \times 4) + 5(2+3)^2 =$

> ### EasyMatch A.4 (C)
> **STAAR – Algebra I: Properties and Attributes of Functions - The student will demonstrate an understanding of the properties and attributes of functions.**
>
> **A.4 – Foundations for functions.** The student understands the importance of the skills required to manipulate symbols in order to solve problems and uses the necessary algebraic skills required to simplify algebraic expressions and solve equations and inequalities in problem situations. The student is expected to
>
> A.4(C) – Connect equation notation with function notation, such as $y = x + 1$ and $f(x) = x + 1$. **Supporting Standard**

STANDARD AND LESSON OVERVIEW
TEACHER DIRECTIONS

A.4(C) – Connect equation notation with function notation, such as $y = x + 1$ and $f(x) = x + 1$. **Supporting Standard**

In connecting equation notation with function notation, it is important note the similarities. Looking at the definition:

Equation – a **mathematical sentence** formed by placing the equal symbol (=) between two expressions;

Function – consists of a set of **input numbers** call the **domain** and a set of **output numbers** called the **range**. There is a pairing of inputs with outputs such that each input is paired with exactly one output.

> The statement that f is a function means that f is a set of ordered pairs $(x, f(x))$, spoken as "x and f of x, or (x, y) such that **no two ordered pairs have the same first element**. A function has two components which are called the Domain and the Range. The **Domain** is the set of all first elements (**x-values**) and the **Range** is the set of all second elements (**y-values**).
>
> Thus, y is also defined as $f(x)$, where the **equation** representing the **functional** relationship between the input and the output, is evaluated for **input x** and

KEEPING IT REAL - ALGEBRA I   A.4(C)

the result is assigned as the respective output or paired **output y**, based on **input x**.

Function Rule: y = x + 5: an equation that describes a function

Given the quadratic equation: $x^2 - 2x - 3 = 0$

Notes

$x^2 - 2x - 3 = 0$	To $y = ax^2 + bx + c$, the standard form
$(x - 3)(x + 1) = 0$	Factoring, results in
$x - 3 = 0 \quad x + 1 = 0$	Set both results equal to 0
$x = 3 \quad x = -1$	**The graph will cross the x axis at (3, 0) and (-1, 0)**
$x = \dfrac{-(-2)}{2(1)}$	To find the vertex, $x = \dfrac{-b}{2a}$
$x = 1$	Using $f(x) = x^2 - 2x - 3$, for the y component
$f(1) = (1)^2 - 2(1) - 3$	f of x, or f of (x = 1)
$f(1) = 1 - 2 - 3$	
$f(1) = -4$	Vertex is (1, -4)

The graph for equation $X^2 - 2x - 3 = 0$, is illustrated below.

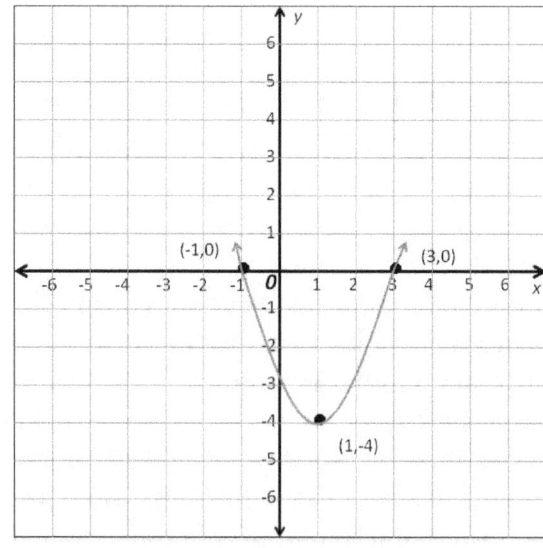

Figure A.4.1

KEEPING IT REAL - ALGEBRA I A.4(C)

The domain is all real numbers and the range is all real numbers for y ≥ -4.

 INTEGRATING TECHNOLOGY

 CULTURAL COMPETENCY

 PROJECT-BASED LEARNING

 WRITING ACROSS THE CURRICULUM

www.keepingitrealeasymatch.com

KEEPING IT REAL - ALGEBRA I   A.4(C)

Student Materials

Objectives for A.4 (C) – Equations and Functional Notation

1. For a function s, s(3) = - 5, and s(-5) = 3. If t = s(x), what is the value of t when x = -5.

_____
_____
_____

2. Evaluate the functions when x = 2, -2 and 4.
   a. $f(x) = 14x + 5$

   b. $f(x) = -6x - 5$

   c. $f(x) = -6.6x$

   d. $f(x) = -3.4x + 15$

   e. $f(x) = 6.3x + 2.8x - 5$

3. Write the equations in function form.
   a. $3x + y = 15$

   b. $5 - y = 16x$

   c. $3x + 5y = 12$

   d. $6y - 12x - 18 = 0$

   e. $2 - 5y - 3 = 9$

Reporting Category 3

EasyMatch A.5 (A)

STAAR – Algebra I: Linear Functions
The student will demonstrate an understanding of linear functions.

A.5 – Linear functions. The student understands that linear functions can be represented in different ways and translates among their various representations. The student is expected to

A.5 (A) – Determine whether or not given situations can be represented by linear functions; **Supporting Standard**

STANDARD AND LESSON OVERVIEW
TEACHER DIRECTIONS

A.5 (A) – Determine whether or not given situations can be represented by linear functions; **Supporting Standard**

Linear functions are used to represent a variety of situations or relationships. The given situation helps to form the mathematical sentence which is composed of the expressions describing the relationships. The sentence is represented by an equation.

A **linear equation** is an equation whose graph is a line. The standard form of a linear equation is:

Ax + By = C, where A, B and C are real numbers and A and B are not zero.

If the domain of a linear function is not specified, it is understood to be all real numbers. The domain is the **x-variable, the independent variable** of the function. The domain can be restricted in the form of a **ray** – a half line extending in one direction.

The **graph** of a line (**linear equation**) is the set of points in a coordinate plane representing all solutions of the equation. Thus, the domain of a linear function when not specified is understood to be all real numbers – all numbers in both of the x-directions, positive (right) and negative (left).

If A is zero in the standard form, the equation becomes **By = C**, or $y = \dfrac{C}{B}$, a constant. Or we could write **y = b**, a constant, **is a function**.

If B is zero, the equation becomes **Ax = C** or $x = \dfrac{C}{A}$. Or we can write, **x = a**, a constant.

Graphs of **y = b** (horizontal line), **a function**, while **x = a** (vertical line), **is Not a function**.

KEEPING IT REAL - ALGEBRA I   A.5(A)

Once the equation is proven to be linear, the equation has to pass the function test to be confirmed as a linear function. Recall the definition of a function,

Function – consists of a set of **input numbers** call the **domain** and a set of **output numbers** called the **range**. There is a pairing of inputs with outputs such that **each input is paired with exactly one output**.

The statement that *f* is a function means that *f* is a set of ordered pairs (x, *f*(x)), spoken as "x and *f* of x," or (x, y) such that **no two ordered pairs have the same first element**. A function has two components which are called the Domain and the Range. The **Domain** is the set of all first elements (**x-values**) and the **Range** is the set of all second elements (**y-values**).

Thus, *y* is also defined as *f*(x), where the **equation** representing the **functional** relationship between the input and the output, is evaluated for **input x** and the result is assigned as the respective output or paired **output y**, based on **input x**.

Function Rule: y = x + 5: an equation that describes a function

Notes: In determining whether a situation can be represented by a linear function:

1. The family of linear functions have the form: $f(x) = mx + b$
2. The parent linear function (most basic function): $f(x) = x$
3. The Function Rule: $y = x + 5$
4. Test; no two ordered pairs have the same first element.
5. Test; each input is paired with exactly one output.
6. The x variable is raised to the first power in linear functions.

 INTEGRATING TECHNOLOGY    CULTURAL COMPETENCY

 PROJECT-BASED LEARNING    WRITING ACROSS THE CURRICULUM

www.keepingitrealeasymatch.com

KEEPING IT REAL - ALGEBRA I

A.5(A)

Student Materials

Objectives for A.5 (A) – Situations Represented by Linear Functions

1. For the data set below, graph and determine whether or not the set represents a linear function:

{(0, 4) , (2, 3), (4, 2), (6, 1), (8, 0)}

Insert your answer below:

2. For Tables A and B below, tell if the situations represent linear functions. Provide discussion or justification for your answer.

Table A

Input	1	2	3	3	5
Output	9	1	2	0	3

Table B

Input	5	6	7	8	9
Output	4	3	2	1	0

KEEPING IT REAL - ALGEBRA I   A.5(A)

3. Graph the two equations and determine if the definition of a linear function is satisfied:
   $y = 5$; $x = 3$;

4. Determine whether the set data pair represented by the table below is discrete or continuous. Is the relationship linear or not?

x	1	1	2	4	5	6	7	8	9	10	11
y	2	3	4	5	5	7	8	7	8	9	10

> ### EasyMatch A.5 (B)
>
> **STAAR – Algebra I: Linear Functions**
> **The student will demonstrate an understanding of linear functions.**
>
> **A.5 – Linear functions.** The student understands that linear functions can be represented in different ways and translates among their various representations. The student is expected to
>
> A.5 (B) – Determine the domain and range for linear functions in given situations;
> **Supporting Standard**

STANDARD AND LESSON OVERVIEW
TEACHER DIRECTIONS

A.5 (B) – Determine the domain and range for linear functions in given situations;
Supporting Standard

Domain and Range for Linear Functions

The **Domain** of the function is the called the **Independent Variable (x)** and the **Range** of the function is called the **Dependent Variable (y)**. The input values for the function are identified **(domain – inputs – x-values)** and then determine the results **(range – outputs – y-values)** for the function. For the function $y = x$; if $x = 1$, then $y = 1$; if $x = 2$, then $y = 2$; if $x = 3$, then $y = 3$; and if $x = 4$, then $y = 4$. Thus, the function $y = x$ consists of the order pairs (0, 0), (1, 1), (2, 2), (3, 3), (4, 4) ... and also extends in the direction of (-1, -1), (-2, -2), (-3, -3) and (-4, -4) ..... For $y = x$, the **Domain (x-values)** is all Real Numbers and the **Range (y-values)** is all Real Numbers.

Considering another approach, a function includes two components which are called the Domain and the Range. The **Domain** is the set of all first elements (**x-values**) and the **Range** is the set of all second elements (**y-values**).

The concepts of a linear function are provided in EasyMatch A.5(A)

KEEPING IT REAL - ALGEBRA I A.5(B)

 INTEGRATING TECHNOLOGY

 CULTURAL COMPETENCY

 PROJECT-BASED LEARNING

 WRITING ACROSS THE CURRICULUM

www.keepingitrealeasymatch.com

KEEPING IT REAL - ALGEBRA I   A.5(B)

Student Materials

Objectives for A.5 (B) – Determine the Domain and Range for the Linear Functions

1. For the data set below, graph and determine the domain and range.

{(0, 4) , (2, 3), (4, 2), (6, 1), (8, 0)}

2. For Tables A and B below, tell if the situations represent linear functions. Determine the domain and range.

Table A

Input	1	2	3	4	5
Output	9	1	2	0	3

Table B

Input	5	6	7	8	9
Output	4	3	2	1	0

KEEPING IT REAL - ALGEBRA I   A.5(B)

3. Determine the domain and range for the function below. Is the relationship linear or not?

x	1	12	2	4	5	6	7	8	9	10	11
y	2	3	4	5	5	7	8	7	8	9	10

_____
_____
_____
_____
_____
_____
_____

4. A small fishing boat is used to help **c** number of campers cross a wide river. The boat trips, **b(t)**, needed to assist the camping group is determined using the function $b(t) = \dfrac{c}{5}$. If there are no more than 20 campers to cross the river, what are the domain and the range of the function for the boat?

_____
_____
_____
_____

5. Identify the domain and range for the function below.

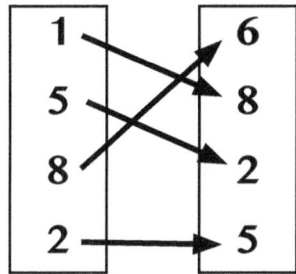

KEEPING IT REAL - ALGEBRA I

A.5(B)

> **EasyMatch A.5 (C)**
>
> **STAAR – Algebra I: Linear Functions**
> **The student will demonstrate an understanding of linear functions.**
> **A.5 – Linear functions.** The student understands that linear functions can be represented in different ways and translates among their various representations. The student is expected to
>
> A.5 (C) – Use, translate, and make connections among algebraic, tabular, graphical, or verbal descriptions of linear functions. **Readiness Standard**

STANDARD AND LESSON OVERVIEW
TEACHER DIRECTIONS

A.5 (C) – Use, translate, and make connections among algebraic, tabular, graphical, or verbal descriptions of linear functions. **Readiness Standard**

Operations can also be performed on a given set of points. A **transformation** applies a rule to the coordinates of the points and produces an image. Some types of transformations are **translations, vertical stretches, vertical shrinks and reflections.**

A **translation** moves all points in a figure the same distance in the same direction either horizontally, vertically, or both. Translations can be described algebraically as described below:

Horizontal translation: $(x, y) \rightarrow (x + h, y)$ **Vertical Translation:** $(x, y) \rightarrow (x, y + k)$

A **vertical stretch** or **shrink** moves every point away from the x-axis (**vertical stretch**) or toward the x-axis (**vertical shrink**), while points on the x-axis remain fixed. A reflection flips a figure in a line. Vertical stretches and shrinks can be described with or without reflection in the x-axis algebraically.

Figure A.5.1 is a **vertical translation** described as $\triangle ABC$, $(x, y) \rightarrow (x, \mathbf{y + 3})$. The y value for each ordered pair was increased by 3. Also in Figure A.5.1 is a reflection with a vertical stretch - $\triangle EFG$. $\triangle E'F'G'$ vertical stretch with reflection on the x-axis in algebraic form: $(x, y) \rightarrow (x, -2y)$.

KEEPING IT REAL - ALGEBRA I   A.5(C)

Original ABC	Transformed A'B'C'	Original EFG	Reflected /Stretched E'F'G'
A(2, 0) →	A'(2, 3)	E(-5, 0) →	E'(-5, 0)
B(4, 0) →	B'(4, 3)	F(-3, 0) →	F'(-3, 0)
C(3, 2) →	C'(3, 5)	G(-4, 2) →	G'(-4, -4)

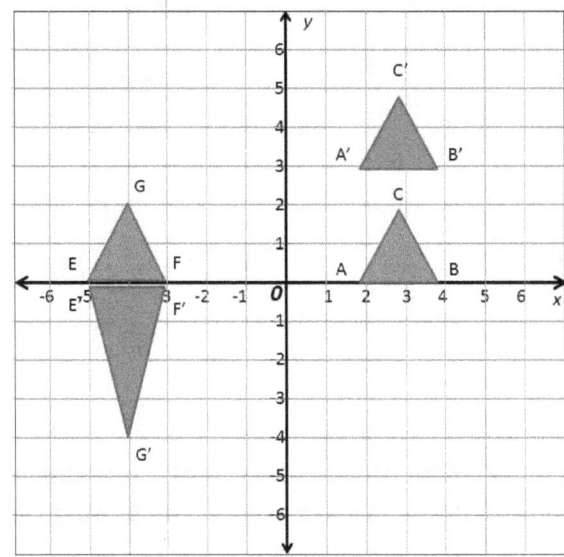

Figure A.5.1

Summary of Transformations:

1. Horizontal Translation: $(x, y) \rightarrow (x + h, y)$

2. Vertical Translation: $(x, y) \rightarrow (x, y+h)$

3. Vertical stretch or shrink without reflection: $(x, y) \rightarrow (x, ay)$, where $a > 0$

4. Vertical stretch or shrink with reflection: $(x, y) \rightarrow (x, ay)$, where $a < 0$

 INTEGRATING TECHNOLOGY    CULTURAL COMPETENCY

 PROJECT-BASED LEARNING    WRITING ACROSS THE CURRICULUM

www.keepingitrealeasymatch.com

KEEPING IT REAL - ALGEBRA I   A.5(C)

Student Materials

Objectives for A.5 (C) – Transformations of Linear Functions

1. Does the translation change the size of the figure? Explain.

_____
_____
_____
_____
_____

2. A rectangle has vertices ABCD at (0, 0), (0, 2), (4, 0), (4, 2). Perform the transformations and draw each figure:

   a. (x, y) → (x-3, y)

   b. (x, y) → (x, y + 3)

   c. (x, y) → (x, 3y)

   d. (x, y) → (x, -2y)

   e. (x, y) → (x + 2, y + 2)

KEEPING IT REAL - ALGEBRA I A.5(C)

3. Graph the following equation: $4x - 6y = 12$.

4. The charge for overdue books at the library is $0.50 per day per book, with a maximum late fee of $5.00 per book. Develop a graph to model the total late fee for 2 books that were checked out on the same day and were overdue.

5. For the function $y = 4 - x$, identify three (3) sets of ordered pairs which are points on the graph of the function.

> # EasyMatch A.6 (A)
> ## STAAR – Algebra I: Linear Functions
> **The student will demonstrate an understanding of linear functions.**
>
> **A.6 - Linear functions.** The student understands the meaning of the slope and intercepts of the graphs of linear functions and zeros of linear functions and interprets and describes the effects of changes in parameters of linear functions in real-world and mathematical situations. The student is expected to
>
> A.6 (A) - Develop the concept of slope as rate of change and determine slopes from graphs, tables, and algebraic representations; **Supporting Standard**

STANDARD AND LESSON OVERVIEW
TEACHER DIRECTIONS

A.6 (A) - Develop the concept of slope as rate of change and determine slopes from graphs, tables, and algebraic representations; **Supporting Standard**

The **slope of a line** is defined as the ratio of the **vertical rise** to the **horizontal run** or the **rate of change** in one quantity to a change in another. We may view a slope as a ramp use to go **up (positive direction)** or **down (negative direction)**. The letter **m** is used to represent the slope of a non-vertical line passing through two points; (x, y) and (x_1, y_1), which is also written as; (x_1, y_1) and (x_2, y_2). Use one of the forms and remain consistent in representing the two points – P_1 and P_2.

The **Point Slope Form, $y - y_1 = m(x - x_1)$**, can be used to find the slope **m**.

Thus, $m = \dfrac{y - y_1}{x - x_1}$ where y = y of P_2 and x = x of P_2. Or $m = \dfrac{y_2 - y_1}{x_2 - x_1}$ for P_1 and P_2.

Example: Let $P_1 (x_1, y_1) = (1, 2)$ and $P_2 (x_2, y_2) = (5, 4)$. Please note the graph in **Figure A.6.1** illustrating the slope of the line passing through the two points – P_1 and P_2. The slope is going up (looks like y = x, the parent function) so the slope is **positive**. The value for **m** is $\dfrac{2}{4} = \dfrac{1}{2}$. Therefore, for every four (4) units the line runs or **changes** to the right, the line will also go up or **change** two (2) units vertically.

KEEPING IT REAL - ALGEBRA I   A.6(A)

Please note: A horizontal line has a slope of zero (0). ↔

A vertical line has an undefined slope. ↕

$$\text{Slope} = \frac{\text{rise}}{\text{run}} = \frac{\text{change in } y}{\text{change in } x} = m = \frac{y_2 - y_1}{x_2 - x_1}$$

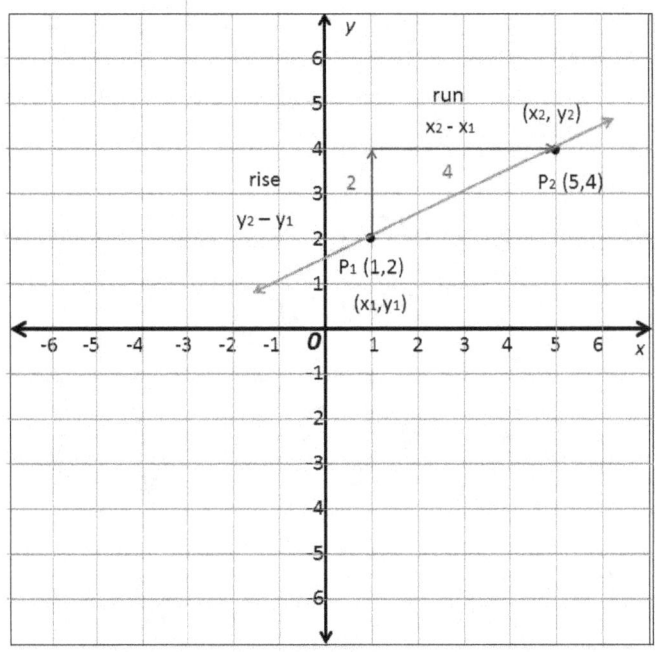

Figure A.6.1 - Slope

 INTEGRATING TECHNOLOGY

 CULTURAL COMPETENCY

 PROJECT-BASED LEARNING

 WRITING ACROSS THE CURRICULUM

www.keepingitrealeasymatch.com

KEEPING IT REAL - ALGEBRA I   A.6(A)

Student Materials

Objectives for A.6 (A) – Slopes and Rate of Change

1. Draw the graphs and find the slopes of the lines passing through the following points.
    a. P_1 (3,4) and P_2 (5,2)
    b. P_1 (1, -4) and P_2 (5, -7)
    c. P_1 (-2, 5) and P_2 (-2, 2)
    d. P_1 (0, -1) and P_2 (7, -4)
    e. P_1 (-3, -4) and P_2 (-2, -1)

2. The cost of using the wireless internet in the local Subway is illustrated in the table below. Find the rate of change in cost with respect to time.

Time in Minutes	10	20	30	40	50
Cost in Dollars	2	4	6	8	10

_____
_____
_____

KEEPING IT REAL - ALGEBRA I A.6(A)

3. After eating a big lunch and lots of cookies, you and your friend go for a daily walk. To keep a consistent heart rate and burn calories, you must walk various distances within the minutes specified in the table below. Find the rate of change in distance with respect to the allotted times.

Time in Minutes	20	40	60
Distance in Miles	1.5	3	4.5

4. Determine whether the slopes of the lines are positive or negative:

   a. (-2, -2) and (2, 2)

   b. (1, 4) and (5, 3)

   c. (0, 2) and (4, 2)

   d. (2, 3) and (8, 0)

> # EasyMatch A.6 (B)
> ## STAAR – Algebra I: Linear Functions
> **The student will demonstrate an understanding of linear functions.**
>
> **A.6 - Linear functions.** The student understands the meaning of the slope and intercepts of the graphs of linear functions and zeros of linear functions and interprets and describes the effects of changes in parameters of linear functions in real-world and mathematical situations. The student is expected to
>
> A.6 (B) - Interpret the meaning of slope and intercepts in situations using data, symbolic representations, or graphs; **Readiness Standard**

STANDARD AND LESSON OVERVIEW
TEACHER DIRECTIONS

A.6 (B) - Interpret the meaning of slope and intercepts in situations using data, symbolic representations, or graphs; **Readiness Standard**

The slope intercept form is one of the basic forms of a linear function with a line as the graph. The following notes were provided earlier in standard A.5 (A).

Notes: In determining whether a situation can be represented by a linear function:

1. The family of linear functions have the form: $f(x) = mx + b$
2. The parent linear function (most basic function): $f(x) = x$
3. The Function Rule: $y = x + 5$

Slope Intercept Form: $y = mx + b$

A **Linear Equation** is an equation of a straight line, with x raised to the power of one and y raised to the power of one. The Domain is all Real Numbers and the Range is all Real Numbers. Given the equation of a line, $2x - 3y = 6$, the first step is to rewrite the equation in the **Slope Intercept Form** which is $y = mx + b$. For the Slope Intercept Form, **m** represents the **slope** and **b** represents the **y-intercept** or the location where the line of the equation crosses the y-axis. Once the equation is in $y = mx + b$ form, set $x = 0$ and then $y = b$, where b represents the point $(0, b)$, the point where the line will intercept or cross the y-axis.

KEEPING IT REAL - ALGEBRA I   A.6(B)

Follow the procedures below.

$2x - 3y = 6$	**Notes**
$-2x + 2x - 3y = -2x + 6$	Using Additive Inverse; add $-2x$ to both sides
$-3y = -2x + 6$	$-2x + 2x = 0$
$-\frac{1}{3}(-3y = -2x + 6)$	Multiplicative Inverse; multiply by $-\frac{1}{3}$:Distributive Property
$y = \frac{2}{3}x - 2$	$m = \frac{2}{3}$, the slope is positive: $b = -2$, the y-intercept for $x = 0$.

For $m = \frac{2}{3}$, the slope is positive, so the function is increasing and the graph will look like the **Identity Function**, $y = x$. The graph of the function $y = \frac{2}{3}x - 2$ is illustrated in **Figure A.6.2** below.

For $x = 0$, $y = -2$ and the y-intercept is at the point **(0,-2)**. The point (0,-2) is the **reference point** which indicates where to start as we plot the slope, $m = \frac{2}{3}$, or $\frac{rise}{run}$, $\frac{y - direction}{x - direction}$.

A common practice is to start at the **reference point (P_1)**, (0, -2) and the denominator (3) indicates to count to the **right** 3 units (run). After counting 3 units to the right, the numerator (2) indicates to count **up** 2 units (slope is **positive**). After plotting P_2 from the reference point, draw a line which passes through both points P_1 and P_2. The slope of the line can be examined for correctness, while the line represents all solution pairs for the given equation.

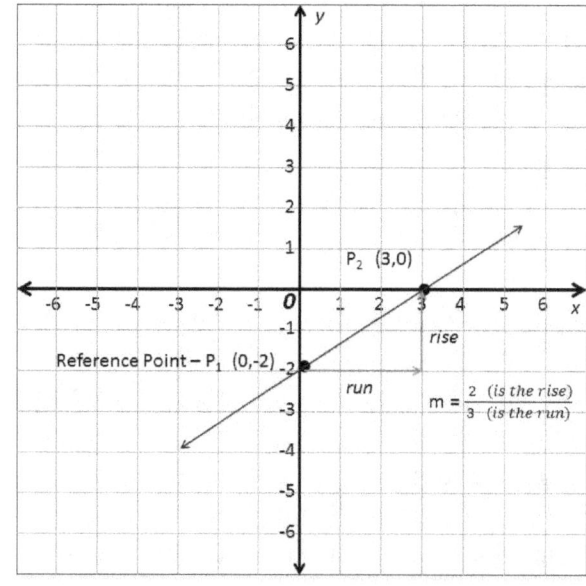

Figure A.6.2 - **Slope Intercept**

KEEPING IT REAL - ALGEBRA I A.6(B)

 INTEGRATING TECHNOLOGY

 CULTURAL COMPETENCY

 PROJECT-BASED LEARNING

 WRITING ACROSS THE CURRICULUM

www.keepingitrealeasymatch.com

KEEPING IT REAL - ALGEBRA I

KEEPING IT REAL - ALGEBRA I

A.6(B)

Student Materials

Objectives for A.6 (B) – Slopes and Intercepts

1. Simplify the equations below to the y-intercept form and graph the line of the solution for the equation.
   - a. $2x + 3y = 6$
   - b. $3x - 4y = 12$
   - c. $3x - 5y = 15$
   - d. $4x - 5y = 20$
   - e. $2x + 3y > 6$
   - f. $3x + 4y \leq 12$
   - g. $2x - 3y > 6$
   - h. $3x - 4y > 12$
   - i. $3x + 5y \geq 15$

KEEPING IT REAL - ALGEBRA I   A.6(B)

2. The table below shows the relationship between the number of dollars a work earns and the number of hours worked. Graph the line which represents the equation. What does the slope of the graph represent?

Time Worked (hours)	4	8	12	16	20
Amount Earned (dollars)	60	120	180	240	300

3. A weight trainer is adding plates of equal weight to a bar. The table below illustrates the total weight - the number of plates of equal weight and the weight for a bar. What is the weight of the bar?

Number of Plates	Total Weight (lbs)
2	110
4	180
6	250

KEEPING IT REAL - ALGEBRA I   A.6(B)

4. Write the equations for the line with the given slope and y-intercept:

    a. Slope: 4 and y-intercept: 2

    b. Slope: -2 and y-intercept: 3

    c. Slope: 3 and y-intercept: -4

> # EasyMatch A.6 (C)
> ## STAAR – Algebra I: Linear Functions
> **The student will demonstrate an understanding of linear functions.**
>
> **A.6 - Linear functions.** The student understands the meaning of the slope and intercepts of the graphs of linear functions and zeros of linear functions and interprets and describes the effects of changes in parameters of linear functions in real-world and mathematical situations. The student is expected to
>
> A.6 (C) – Investigate, describe, and predict the effects of changes in m and b on the graph of y = mx + b; **Readiness Standard**

STANDARD AND LESSON OVERVIEW
TEACHER DIRECTIONS

A.6 (C) – Investigate, describe, and predict the effects of changes in m and b on the graph of y = mx + b; **Readiness Standard**

On the graph of the line for y = mx + b, it is important to understand each component or variable of the equation and the effects of changes. For example, changes in the ratio of the rise (y-axis component) versus the run (x-axis component) will impact the slope. If b, the y-intercept, is increased or decreased, the y-intercept point will move up (increase) or down (decrease) on the y-axis. Consider **Figure A.6.3**, with a slope of $m_1 = \frac{2}{3}$, and a y-intercept of b = -2, **Reference Point 1**, for the line connecting points $P_1(0, -2)$ and $P_2(3,0)$. If the slope were changed from $m_1 = \frac{2}{3}$, to $m_2 = \frac{5}{3}$, the increase in the rise from 2 to 5 would change the slope and likewise, the line would have an increased slope. If the value of b, the y-intercept, were changed from a -2 to a positive 1, the y-intercept would be at a new **Reference Point 2** located at $P_3(0, 1)$. With $m_2 = \frac{5}{3}$, the run of 3 to the right from the **Reference Point 2** and the rise of 5 units up places P_4 at (3,6). **Figure A.6.3** provides a simple example for the effects of changes in **m** and **b** and allows for further predictions or investigations based on

KEEPING IT REAL - ALGEBRA I A.6(C)

selected changes in the **m** and **b** values.

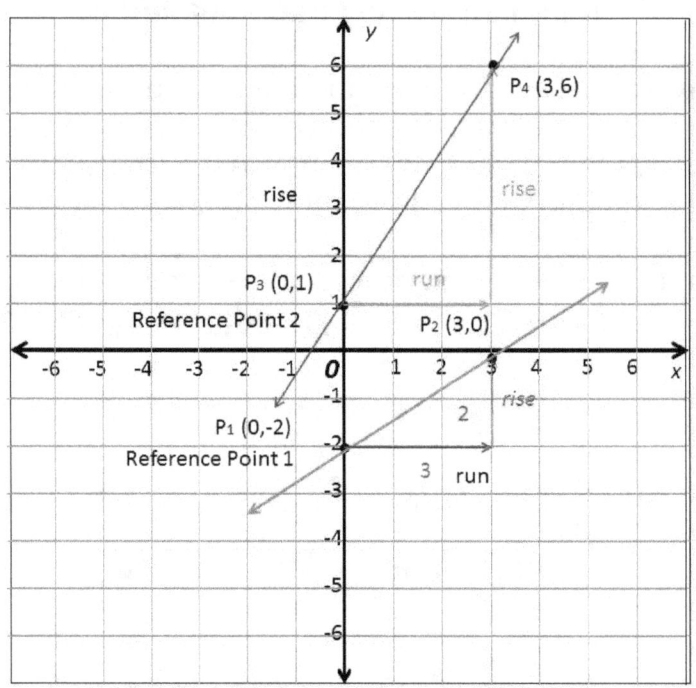

Figure A.6.3 - Effects of Changes

KEEPING IT REAL - ALGEBRA I

A.6(C)

Student Materials

Objectives for A.6 (C) – Effects of Changes in m and b

1. Graph two lines f and g based on the following description. The graph for line g is determined from line f by multiplying the slope of line f by a -2 and decreasing the y-intercept of line f by 5. Label the points for the y-intercepts, and illustrate the slopes by labeling the rise and run for each line.

2. The graph of line r represents $y = \frac{1}{5}x - 1$. If the slope of r is multiplied by a -5 to create line p, what true statement can be made about the graphs of the two lines?

_____
_____
_____
_____
_____
_____

KEEPING IT REAL - ALGEBRA I   A.6(C)

3. The graph of y = 3x + 2 is translated 2 units up. Write the equation which describes the line of the new graph. Draw the line for the new equation.

_____

4. Identify the slopes and y-intercepts for each line of the given equations. Graph each line for the equations.

    a. y = 3x + 2

    b. y = -x + 1

    c. y = 6 + 3x

    d. $y = \frac{1}{4}x + 8$

    e. $y = \frac{2}{3}x - 1$

KEEPING IT REAL - ALGEBRA I A.6(C)

5. For the equations in problem 4, multiply the slopes by -1 and add 2 to the y- intercept. Graph the lines for the modified equations

KEEPING IT REAL - ALGEBRA I

A.6(C)

> # EasyMatch A.6 (D)
>
> ## STAAR – Algebra I: Linear Functions
> **The student will demonstrate an understanding of linear functions.**
>
> **A.6 - Linear functions.** The student understands the meaning of the slope and intercepts of the graphs of linear functions and zeros of linear functions and interprets and describes the effects of changes in parameters of linear functions in real-world and mathematical situations. The student is expected to
>
> A.6 (D) - Graph and write equations of lines given characteristics such as two points, a point and a slope, or a slope and y-intercept; **Supporting Standard**

STANDARD AND LESSON OVERVIEW
TEACHER DIRECTIONS

A.6 (D) - Graph and write equations of lines given characteristics such as two points, a point and a slope, or a slope and y-intercept; **Supporting Standard**

There are various approaches to graphing and writing equations for lines given certain characteristics. Some common approaches are:

1. Given two points, draw a line through the two points
2. Given one point and a slope, plot the second point and draw a line through the two points.
3. Given a y-intercept and a slope, plot the second point and draw a line through the two points.
4. Determine the zeros of linear functions.
5. Equations can be written from graphs of lines based on the given characteristics.

Given Two Points, Draw a Line through the Two Points

One can use **Figure A.6.3 – Effects of Changes**, which has the graph of two lines to illustrate the concepts. The graph of the upper line is drawn through points P_3 (0,1) and P_4 (3,6) and the graph of the lower line is drawn through points P_1 (0,-2) and P_2 (3,0). From the graphs of each line, one can determine the slope, **m**, and the y-intercept, **b**, for each line. Going back to the form of a linear equation for a straight line **y = mx + b**, the equations for the graphs of the lines are:

KEEPING IT REAL - ALGEBRA I   A.6(D)

Upper Line: $y = \dfrac{5}{3}x + 1$ **Lower Line:** $y = \dfrac{2}{3}x - 2$

Given One Point or the Y-intercept Point and a Slope

Consider the upper line with the point P_3 (0,1) and the slope of $m = \dfrac{5}{3}$. One can draw the graph of the line and determine the equation. Using the point P_3 (0,1), the y-intercept as the Reference Point, which indicates where to start as we plot the slope, $m = \dfrac{5}{3}$, or $\dfrac{rise}{run}$, $\dfrac{y-direction}{x-direction}$. A common practice used to plot the slope is to start at the **Reference Point** P_3 **(0,1)** and use the value of the denominator (3) which indicates to count to the **right** 3 units (run) and stop. After counting 3 units to the right, the numerator (5) indicates to count **up** 5 units (slope is **positive**) and stop. After counting 3 units to the right and then 5 units up, plot the point of the current location P_4 **(3,6)**. From the reference point, draw a line which passes through both points P_3 and P_4. The equation which represents the graph of the upper line is: $y = \dfrac{5}{3}x + 1$.

Zero of Linear Functions

A zero of a function **y = f(x)** is an x-value for which **f(x) = 0** (or **y = 0**). **Because y = 0 along the x-axis of the coordinate plane, a zero of a function is an x-intercept of the function's graph.** Consider the equation, f(x) = 2x – 3.

Notes

f(x) = 2x – 3	Next step is to substitute f(x) with zero (0)
2x – 3 = 0	Both sides are equal, exchange sides of equation
2x – 3 + 3 = 0 + 3	Additive Inverse, add 3 to both sides
2x = 3	-3 + 3 = 0
$\dfrac{1}{2}(2x = 3)$	Multiplicative Inverse to make the coefficient of x equal to one (1)
$x = \dfrac{3}{2}$	The zero of **f(x) = 2x – 3** is $\dfrac{3}{2}$

KEEPING IT REAL - ALGEBRA I   A.6(D)

For the upper line in **Figure A.6.3**:

$$y = \frac{5}{3}x + 1$$ Next step is to substitute y with zero (0)

$$\frac{5}{3}x + 1 = 0$$ Both sides are equal, exchange sides of equation

$$\frac{5}{3}x + 1 - 1 = 0 - 1$$ Additive Inverse, subtracting 1 from both sides

$$\frac{5}{3}x = -1$$ Next, applying the Multiplicative Inverse

$$\frac{3}{5}(\frac{5}{3}x = -1)$$ Multiplicative Inverse to make coefficient of x equal to one (1)

$$x = -\frac{3}{5}$$ The zero of $y = \frac{5}{3}x + 1$ is $-\frac{3}{5}$

Given the characteristics of lines (points, slopes and y-intercepts) one can graph the lines and write the equations representing the respective lines and determine the zero of linear functions.

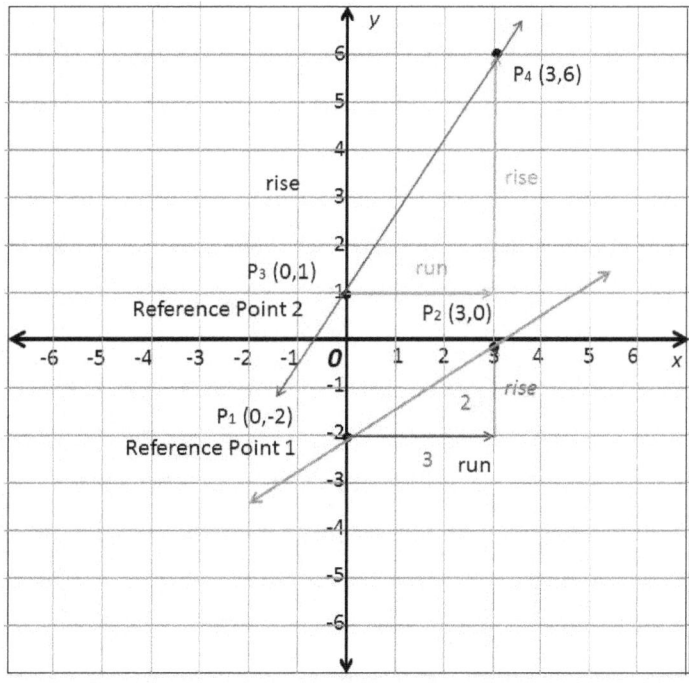

Figure A.6.3 - Effects of Changes

117

KEEPING IT REAL - ALGEBRA I A.6(D)

 INTEGRATING TECHNOLOGY

 CULTURAL COMPETENCY

 PROJECT-BASED LEARNING

 WRITING ACROSS THE CURRICULUM

www.keepingitrealeasymatch.com

KEEPING IT REAL - ALGEBRA I   A.6(D)

Student Materials

Objectives for A.6 (D): Graph and Write Equations of Lines

1. Write the equations of the lines that pass through the given points.
    a. (-2,-1), (2,1)
    b. (0,2), (3,0)
    c. (4,5), (0,3)

2. Write an equation of the line that passes through the give point and has slope m.
    a. (-1, 3); m = 3
    b. (5, 3); m = -2
    c. (2, -3); m = 4

3. Write the equation of the line with the given slope and the y-intercept.

    a. Slope: 3; y-intercept; 3
    b. Slope: -2; y-intercept; 5
    c. Slope: 1; y-intercept; -3

KEEPING IT REAL - ALGEBRA I   A.6(D)

4. Determine the zero of the function represented by the graph of the lower line in **Figure A.6.3**.

EasyMatch A.6 (E)

STAAR – Algebra I: Linear Functions
The student will demonstrate an understanding of linear functions.

A.6 - Linear functions. The student understands the meaning of the slope and intercepts of the graphs of linear functions and zeros of linear functions and interprets and describes the effects of changes in parameters of linear functions in real-world and mathematical situations. The student is expected to

A.6 (E) - Determine the intercepts of the graphs of linear functions and zeros of linear functions from graphs, tables, and algebraic representations; <u>**Supporting Standard**</u>

STANDARD AND LESSON OVERVIEW
TEACHER DIRECTIONS

A.6 (E) - Determine the intercepts of the graphs of linear functions and zeros of linear functions from graphs, tables, and algebraic representations; <u>**Supporting Standard**</u>

The slope intercept form is one of the basic forms of a linear function with a line as the graph. The following notes were provided earlier in standard A.5 (A).

Notes: In determining whether a situation can be represented by a linear function:

1. The family of linear functions have the form: $f(x) = mx + b$
2. The parent linear function (most basic function): $f(x) = x$
3. The Function Rule: $y = x + 5$

Slope Intercept Form: $y = mx + b$

A **Linear Equation** is an equation of a straight line, with x raised to the power of one and y raised to the power of one. The Domain is all Real Numbers and the Range is all Real Numbers. Given the equation of a line, **$2x - 3y = 6$,** the first step is to rewrite the equation in the **Slope Intercept Form** which is **$y = mx + b$**. For the Slope Intercept Form, **m** represents the **slope** and **b** represents the **y-intercept** or the location where the line of the equation crosses the y-axis. Once the equation is in $y = mx + b$ form, set $x = 0$ and then $y = b$, where b represents the point **(0, b)**, the point where the line will intercept or cross the y-axis. Follow the procedures below.

KEEPING IT REAL - ALGEBRA I A.6(E)

2x - 3y = 6   Notes

-2x + 2x - 3y = -2x + 6   Using Additive Inverse; add -2x to both sides
      -3y = -2x + 6   -2x + 2x = 0

$-\frac{1}{3}(-3y = -2x + 6)$ Multiplicative Inverse; multiply by $-\frac{1}{3}$: Distributive Property

$y = \frac{2}{3}x - 2$ $m = \frac{2}{3}$, the slope is positive: b = -2, the y-intercept for x = 0.

For $m = \frac{2}{3}$, the slope is positive, so the function is increasing and the graph will look like the **Identity Function**, y = x. The graph of the line for the function $y = \frac{2}{3}x - 2$ is illustrated in **Figure A.6.2** which is illustrated again below.

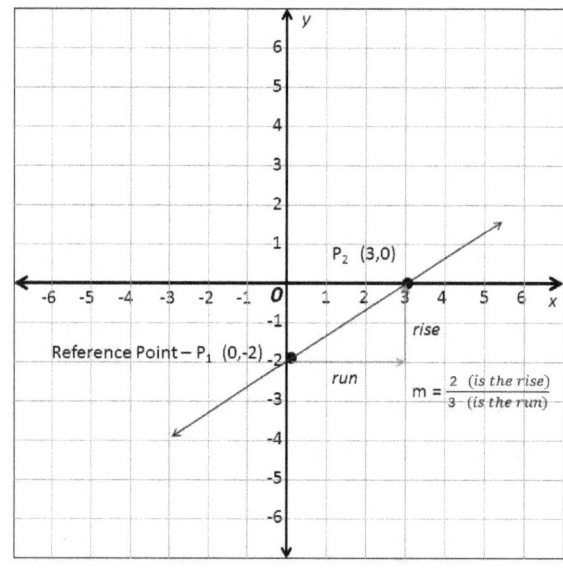

Figure A.6.2

Note that the y-intercept point of $P_1(0,-2)$ is also referred to as the Reference Point. The Reference Point is the starting point from which the slope of $m = \frac{2}{3}$ will be applied to plot point $P_2(3, 0)$. The line drawn through the points P_1 and P_2 represents the graph of the line for the equation $y = \frac{2}{3}x - 2$

KEEPING IT REAL - ALGEBRA I A.6(E)

Zero of a Linear Function

The zero of a function y = f(x) is an x-value for which f(x) = 0 (or y = 0). Because y = 0 along the x-axis of the coordinate plane, a zero of a function is an x-intercept of the function's graph. Consider the equation, f(x) = 2x − 3.

Notes

f(x) = 2x − 3	Next step is to substitute f(x) with zero (0)
2x − 3 = 0	Both sides are equal, exchange each side
2x − 3 + 3 = 0 + 3	Additive Inverse, add 3 to both sides
2x = 3	
$\frac{1}{2}(2x = 3)$	Multiplicative Inverse to make the coefficient of x equal to one (1)
$x = \frac{3}{2}$	The zero of the function, **f(x) = 2x − 3** is $\frac{3}{2}$

Going back to the graph of the line for the function $y = \frac{2}{3}x - 2$ as illustrated in **Figure A.6.2**, determine the zero of the function.

$y = \frac{2}{3}x - 2$	Next step is to substitute y with zero (0)
$\frac{2}{3}x - 2 = 0$	Both sides are equal, exchange each side
$\frac{2}{3}x - 2 + 2 = 0 + 2$	Additive Inverse, adding 2 to both sides
$\frac{2}{3}x = 2$	Next, applying the Multiplicative Inverse
$\frac{3}{2}(\frac{2}{3}x = 2)$	Multiplicative Inverse to make coefficient of x equal to one (1)
x = 3	The zero for the function $y = \frac{2}{3}x - 2$ **is 3**.

As illustrated in Figure A.6.2, the x-intercept is at point **P$_2$ (3,0)**.

KEEPING IT REAL - ALGEBRA I   A.6(E)

 INTEGRATING TECHNOLOGY

 CULTURAL COMPETENCY

 PROJECT-BASED LEARNING

 WRITING ACROSS THE CURRICULUM

www.keepingitrealeasymatch.com

KEEPING IT REAL - ALGEBRA I   A.6(E)

Student Materials

Objectives for A.6 (E): Determine the Y-intercept and Zero of a Function

1. What is the zero of the linear function graphed below?

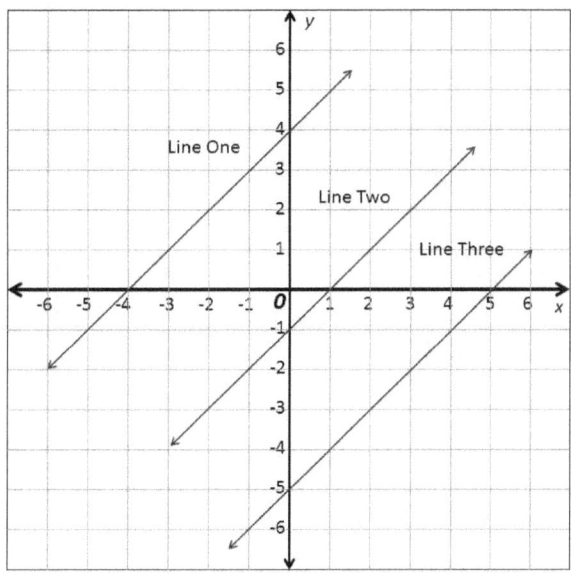

_____
_____
_____
_____

2. Determine the y-intercept for each function represented by lines one, two and three in problem 1 above.

_____
_____
_____

3. Write the equation, in **y = mx + b** form, for each line in problem 1 above.

_____
_____
_____

KEEPING IT REAL - ALGEBRA I   A.6(E)

4. Determine the y-intercept and the zero of the function for each function represented in the tables below.

Table One

Points on the x-axis	0	2	4	6
Points on the y-axis	6	3	0	-3

___
___
___

Table Two

Points on the x-axis	-4	-2	2	4	6
Points on the y-axis	-2	-1	1	2	3

___
___
___

5. Determine the y-intercept and the zero of the function for each function represented in the tables below.

    a. $3x + 4y \leq 12$

    b. $2x - 3y > 6$

    c. $3x - 4y > 12$

    d. $3x + 5y \geq 15$

EasyMatch A.6 (F)

STAAR – Algebra I: Linear Functions
The student will demonstrate an understanding of linear functions.

A.6 - Linear functions. The student understands the meaning of the slope and intercepts of the graphs of linear functions and zeros of linear functions and interprets and describes the effects of changes in parameters of linear functions in real-world and mathematical situations. The student is expected to

A.6 (F) - Interpret and predict the effects of changing slope and y-intercept in applied situations; **Readiness Standard**

STANDARD AND LESSON OVERVIEW
TEACHER DIRECTIONS

A.6 (F) - Interpret and predict the effects of changing slope and y-intercept in applied situations; **Readiness Standard**

On the graph of the line for y = mx + b, it is important to understand each component or variable of the equation and the effects of changes. For example, changes in the ratio of the rise (y-axis component) versus the run (x-axis component) will impact the slope. If b, the y-intercept, is increased or decreased, the y-intercept point will move up (increase) or down (decrease) on the y-axis. Consider **Figure A.6.3** below, with a slope of $m_1 = \frac{2}{3}$, and a y-intercept of **b = -2, Reference Point 1**, for the line connecting points $P_1(0, -2)$ and $P_2(3,0)$. If the slope were changed from $m_1 = \frac{2}{3}$, to $m_2 = \frac{5}{3}$, the increase in the rise from 2 to 5 would change the slope and likewise, the line would have an increased slope. If the value of **b**, the y-intercept, were changed from a -2 to a positive 1, the y-intercept would be at a new **Reference Point 2** located at $P_3(0, 1)$. With $m_2 = \frac{5}{3}$, the run of 3 to the right from the **Reference Point 2** and the rise of 5 units up places P_4 at (3,6). **Figure A.6.3** provides a simple example for the effects of changes in **m** and **b** and allows for further predictions or investigations based on selected

KEEPING IT REAL - ALGEBRA I   A.6(F)

changes in the **m** and **b** values.

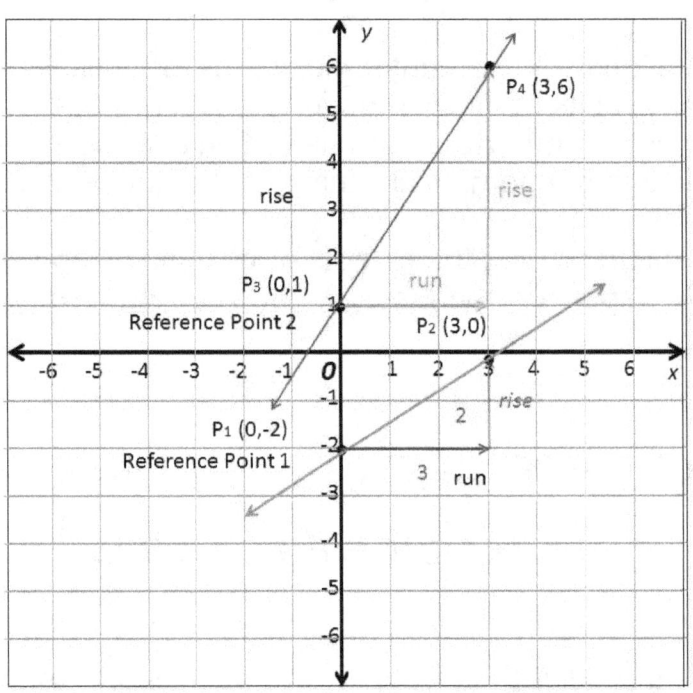

Figure A.6.3 - Effects of Changes

KEEPING IT REAL - ALGEBRA I A.6(F)

Student Materials

Objectives for A.6 (F): Determine the Effects of changing the slope and Y-intercept.

1. A school fundraiser will sell backpacks to raise money. There were some backpacks left over from the previous year, and 40 boxes of backpacks will be ordered this year. When the order arrives, the total number of backpacks for the fundraiser can be determined using the function **f(x) = 48x + 37**, where x represents the number of boxes ordered. If the number of backpacks per box changes so that the situation is modeled by the function **g(x) = 24x + 37**, then how many fewer backpacks will the school have available to sell if they still order 40 boxes?

_____
_____
_____
_____
_____
_____

2. A fighter jet on an aircraft carrier starts its descent for landing based on **h(x) = -100x + 20,000**, where x represents the horizontal distance in miles from where the jet begins its descent. Base on weather conditions, the jet's descent will be modeled by the function **j(x) = -100x + 21,000**. Write a statement which describes the change.

_____
_____
_____
_____
_____
_____

KEEPING IT REAL - ALGEBRA I   A.6(F)

3. Fuel is being pumped at a constant rated represented by a line which passes through the points (1, 10) and (4, 4). The height of the fuel level is represented in inches and the time in minutes. If the rate in which the fuel is pumped is changed by 4 inches per minute and the initial fuel level stays the same, how would the time to empty the tank change?

_____
_____
_____
_____
_____
_____

4. Graph the lines for each equation below. For the equations below, add 2 to the b and multiply the m by a -1 and then graph and compare the modified equations. What are the effects of the change in slope and y-intercept?

a. $2x + 3y = 6$

b. $3x - 4y = 12$

c. $3x - 5y = 15$

d. $4x - 5y = 20$

_____
_____
_____
_____

EasyMatch A.6 (G)

STAAR – Algebra I: Linear Functions
The student will demonstrate an understanding of linear functions.

A.6 - Linear functions. The student understands the meaning of the slope and intercepts of the graphs of linear functions and zeros of linear functions and interprets and describes the effects of changes in parameters of linear functions in real-world and mathematical situations. The student is expected to

A.6 (G) - Relate direct variation to linear functions and solve problems involving proportional change. **Supporting Standard**

STANDARD AND LESSON OVERVIEW
TEACHER DIRECTIONS

A.6 (G) - Relate direct variation to linear functions and solve problems involving proportional change. **Supporting Standard**

The concept of proportional change is often encountered when the dimensions of a two-dimensional object or figure are changed. Such change will affect the perimeter and area. When the figure is enlarged or reduced, a new figure is created which is similar to the original figure. The **scale factor** is used to represent the ratio of the lengths of the two corresponding sides for the similar figures. If two figures are similar with a scale factor of a:b, then the following is true:

1. The ratio of the perimeters of the figures is a:b.
2. The ratio of the areas of the figures is $a^2 : b^2$

In Figure A.6.4, Object ABCD was changed to Object A'B'C'D' and Object EFG was changed to Object E'F'G'.

KEEPING IT REAL - ALGEBRA I A.6(G)

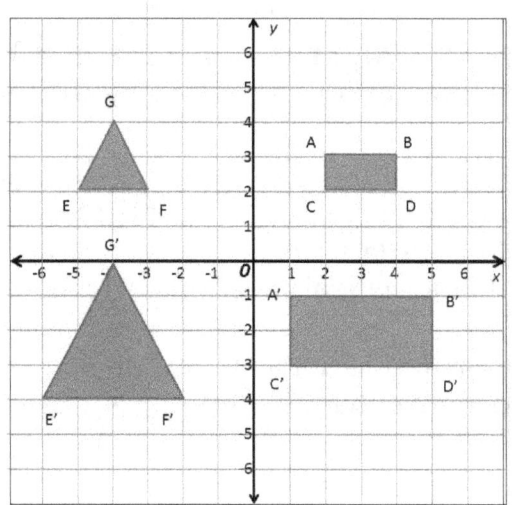

Figure A.6.4

For the objects in Figure A.6.4, the scale factor for the corresponding sides can be determined as follows: The dimensions of the sides in Object EFG were doubled to create Object E'F'G'. Let x be the length of a side of EFG. Then 2x is the length of the corresponding side E'F'G'. The ratio of the lengths of these sides is:

$\dfrac{x}{2x} = \dfrac{1}{2}$, or 1:2 The area of Object EFG is 2 square units. To find the area of E'F'G',

$\dfrac{1^2}{2^2} = \dfrac{2}{y}$, when simplified, $\dfrac{1}{4} = \dfrac{2}{y}$.

$1 \times y = 2 \times 4$ products property and simplify

y = 8 **The area of E'F'G' is 8 square units.**

 INTEGRATING TECHNOLOGY    CULTURAL COMPETENCY

 PROJECT-BASED LEARNING    WRITING ACROSS THE CURRICULUM

www.keepingitrealeasymatch.com

KEEPING IT REAL - ALGEBRA I A.6(G)

Student Materials

Objectives for A.6 (G): Proportional Change.

1. Find the area of Object A'B'C'D ' based on the ratio of 1:2 in Figure A.6.4.

2. The value of h varies directly with t. Write a function which represents the relationship between t and h, if $h = \dfrac{10}{3}$ when t = 20.

_____
_____
_____
_____

3. Describe the effect of the area of a circle when the area is halved. That is the scale factor of the original circle to the new circle is $1 : \dfrac{1}{2}$

_____
_____
_____
_____

KEEPING IT REAL - ALGEBRA I

A.6(G)

4. The dimensions of a hexagon are doubled to create a larger hexagon. The perimeter of the smaller hexagon is 20 inches. What is the perimeter of the larger hexagon?

___

5. A rectangular picture is enlarged so that its dimensions are 3 times as wide and 3 times as long as the original picture. The area of the enlarged picture is 135 square inches. What is the area of the original picture?

Reporting Category 4

EasyMatch A.7 (A)

STAAR – Algebra I: Linear Equations and Inequalities.
The student will formulate and use linear equations and inequalities.

A.7 -Linear functions. The student formulates equations and inequalities based on linear functions, uses a variety of methods to solve them, and analyzes the solutions in terms of the situation. The student is expected to

A.7 (A) - Analyze situations involving linear functions and formulate linear equations or inequalities to solve problems; **Supporting Standard**

STANDARD AND LESSON OVERVIEW
TEACHER DIRECTIONS

A.7 (A) - Analyze situations involving linear functions and formulate linear equations or inequalities to solve problems; **Supporting Standard**

A **Linear Equation** is an equation of a straight line, with x raised to the power of one and y raised to the power of one. The Domain is all Real Numbers and the Range is all Real Numbers. A **linear inequality** in two variables, such as x - 2y < 4, is the result of replacing the = sign in a linear equation with $<, \leq, >$ or $\geq$. A solution of an inequality in two variables x and y is an ordered pair (x, y) that produces a true statement when the values of x and y are substituted into the inequality. Given the inequality $2x + 4y \geq 12$, rewrite the inequality in Slope Intercept form, **y = mx + b,** and graph.

$2x + 4y \geq 12$	Notes
$-2x + 2x + 4y \geq -2x + 12$	Additive Inverse; add -2x to both sides
$\frac{1}{4}(4y \geq -2x + 12)$	Multiplicative Inverse, multiply by $\frac{1}{4}$ and Distributive Property
$y \geq -\frac{1}{2}x + 3$	Solution in y-intercept form.

The line is $y = -\frac{1}{2}x + 3$ and the Slope Intercept form is $y \geq -\frac{1}{2}x + 3$, so the line is a **solid line** and the $\geq$ requires that the **area above the line** is shaded. The y-intercept is the point (0, 3) and the slope m is $-\frac{1}{2}$.

KEEPING IT REAL - ALGEBRA I — A.7(A)

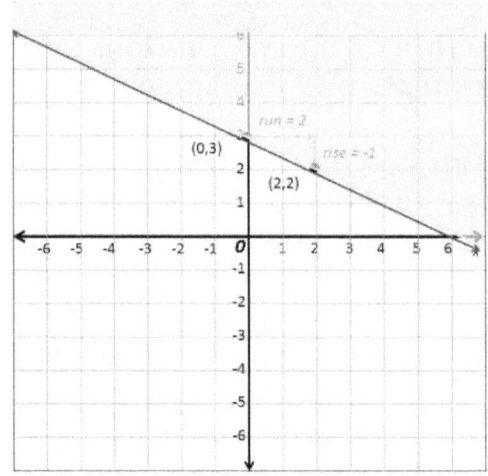

Figure A.7.1

Given the inequality 2x - 3y > 6, rewrite the inequality in the Slope Intercept form and graph.

2x - 3y > 6	Notes
- 2x + 2x - 3y > - 2x + 6	Additive Inverse; add -2x to both sides
$-\frac{1}{3}(-3y < -2x + 6)$	Multiplicative Inverse; multiply by $-\frac{1}{3}$, which requires to change > to <
$y < \frac{2}{3}x - 2$	**Note: When multiplying an inequality by a negative value, change the direction of the inequality.**

Since the line is $y = \frac{2}{3}x - 2$ and the Slope Intercept form is $y < \frac{2}{3}x - 2$, the line cannot be included. Thus, the line must be dashed and shade the area below the dashed line, since the inequality is less than. The graph for the inequality $y < \frac{2}{3}x - 2$ is illustrated in Figure A.7.2.

KEEPING IT REAL - ALGEBRA I

A.7(A)

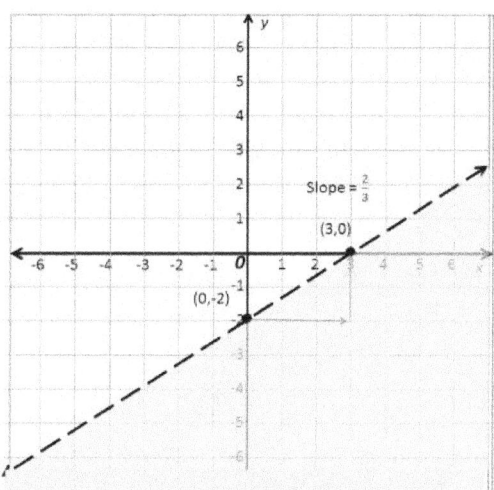

Figure A.7.2

 INTEGRATING TECHNOLOGY

 CULTURAL COMPETENCY

 PROJECT-BASED LEARNING

 WRITING ACROSS THE CURRICULUM

www.keepingitrealeasymatch.com

KEEPING IT REAL - ALGEBRA I

A.7(A)

Student Materials

Objectives for A.7 (A): Linear Equations and Inequalities

1. Given the inequalities below, rewrite in slope intercept form and graph.
    a. $3x + 4y \leq 8$
    b. $2x - 3y > 9$
    c. $3x - 2y > 6$
    d. $2x + 3y \geq 12$

2. An architect designed a hospital with f floors that will have an antenna 16 m tall on its roof. Each floor of the hospital will be 4 m high. Write a function which can be used to find the total height of the hospital in meters, including the antenna.

_____
_____
_____
_____

KEEPING IT REAL - ALGEBRA I   A.7(A)

3. Write inequalities for the following situations:

   a. A child must be 48 inches tall to ride the boat.

   b. A bus can seat 45 passengers of less.

   c. The total number of boys and girls for the game is 15 or less.

4. The local bank offers free checking for accounts with a of $300 or more. Suppose you have a balance of $350 dollars and you just wrote a check for $75 dollars to purchase a game. How much must you deposit to avoid being charged a service fee?

_____
_____
_____
_____

5. In order to receive and A in your English class, you must earn 500 points of reading credit. You have already earned 375 points. Write an inequality and solve to determine the minimum number of points of reading credit you must still earn to receive an A in English.

_____
_____
_____
_____

KEEPING IT REAL - ALGEBRA I   A.7(A)

EasyMatch A.7 (B)

STAAR – Algebra I: Linear Equations and Inequalities.
The student will formulate and use linear equations and inequalities.

A.7 -Linear functions. The student formulates equations and inequalities based on linear functions, uses a variety of methods to solve them, and analyzes the solutions in terms of the situation. The student is expected to

A.7 (B) - Investigate methods for solving linear equations and inequalities using [concrete] models, graphs, and the properties of equality, select a method, and solve the equations and inequalities; **Supporting Standard**

STANDARD AND LESSON OVERVIEW
TEACHER DIRECTIONS

A.7 (B) - Investigate methods for solving linear equations and inequalities using [concrete] models, graphs, and the properties of equality, select a method, and solve the equations and inequalities; **Supporting Standard**

A **linear inequality** in two variables, such as x - 2y < 4, is the result of replacing the = sign in a linear equation with <, ≤, > or ≥. A solution of an inequality in two variables x and y is an ordered pair (x, y) that produces a true statement when the values of x and y are substituted into the inequality.

In solving inequalities graphically, write the inequality in one of the following forms:

1. The forms are: ax + b < 0, ax + b ≤ 0, ax + b > 0 or ax + b ≥ 0.
2. Write the related equation: y = ax + b.
3. Graph the equation y = ax + b.

Using an example:

3x + 3 > 6	**Notes**
3x - 3 > 0	Additive Inverse, subtract 6 from both sides
y = 3x – 3	Writing the related equation
y = 3x – 3	Graph the equation
0 = 3x – 3	Find the x-intercept
3 = 3x	Additive Inverse, add 3 to both sides

KEEPING IT REAL - ALGEBRA I   A.7(B)

x = 1   x-intercept of the graph
y = -3   y – intercept of the graph, where x = 0.

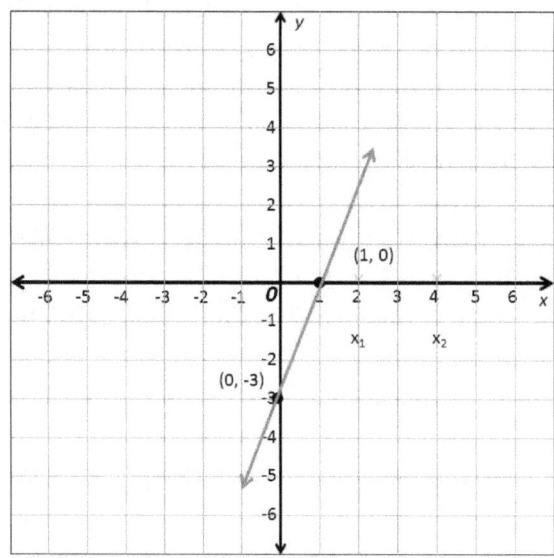

Figure A.7.3

The solutions are all real numbers greater than 1. Check by substituting a number greater than 1 in the original equation.

     3x + 3 > 6   Original Equation

     3(2) + 3 > 6   Substituting 2 in the original equation

     9 > 6   Solution checks.

 INTEGRATING TECHNOLOGY    CULTURAL COMPETENCY

 PROJECT-BASED LEARNING    WRITING ACROSS THE CURRICULUM

www.keepingitrealeasymatch.com

KEEPING IT REAL - ALGEBRA I

A.7(B)

Student Materials

Objectives for A.7 (B): Using Graphs to Solve Linear Inequalities

1. Solve the inequalities below graphically.
    a. $2x + 4 > 8$
    b. $3x + 6 \leq 12$
    c. $5x - 15 < 30$

2. Your cell phone plan costs $50 per month for a given number of minutes. Each additional minute or part of a minute costs $0.25. Your budget is $65 dollars per month for phone costs. What are the possible additional minutes x that you can afford each month?

_____
_____
_____
_____

KEEPING IT REAL - ALGEBRA I   A.7(B)

3. A mechanic charges $65 per hour for labor plus a $50 diagnostic computer fee. The customer has to buy the parts. The total charge for the mechanic was $1,000. How many hours did the mechanic work?

_____

_____

_____

_____

4. Graph the equation $y < 2 - 4x$. Identify three coordinate pairs in the solution set.

5. If $y = \dfrac{3}{4}x - 3$, what is the value of x when y = -5?

EasyMatch A.7 (C)

**STAAR – Algebra I: Linear Equations and Inequalities.
The student will formulate and use linear equations and inequalities.**

A.7 -Linear functions. The student formulates equations and inequalities based on linear functions, uses a variety of methods to solve them, and analyzes the solutions in terms of the situation. The student is expected to

A.7 (C) - Interpret and determine the reasonableness of solutions to linear equations and inequalities. **Supporting Standard**

STANDARD AND LESSON OVERVIEW
TEACHER DIRECTIONS

A.7 (C) - Interpret and determine the reasonableness of solutions to linear equations and inequalities. **Supporting Standard**

An inequality is a mathematical sentence formed by placing one of the symbols $<$, $\leq$, $>$ or $\geq$ between two expressions. On a number line, the graph of an inequality in one variable is the set of points that represent all solutions of the inequality. The graph of an inequality in one variable, uses an open circle for $<$ or $>$ and a closed circle for $\leq$ or $\geq$. The graphs of $x < 2$ and $x \geq 4$ are shown below.

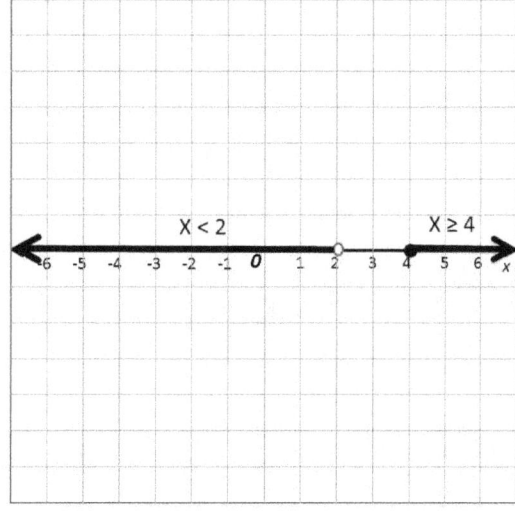

Figure A.7.4

KEEPING IT REAL - ALGEBRA I   A.7(C)

A **linear inequality** in two variables, such as 2x + 4y ≥ 12, is the result of replacing the = sign in a linear equation with ≥. A solution of an inequality in two variables x and y is an ordered pair (x, y) that produces a true statement when the values of x and y are substituted into the inequality. Given the inequality 2x + 4y ≥ 12, rewrite the inequality in Slope Intercept form, **y = mx + b**, and graph.

$$2x + 4y \geq 12 \qquad \text{Notes}$$

$$-2x + 2x + 4y \geq -2x + 12 \quad \text{Additive Inverse; add -2x to both sides}$$

$$\frac{1}{4}(4y \geq -2x + 12) \quad \text{Multiplicative Inverse, multiply by } \frac{1}{4} \text{ and Distributive Property}$$

$$y \geq -\frac{1}{2}x + 3 \quad \text{Solution in y-intercept form.}$$

The line is $y = -\frac{1}{2}x + 3$ and the Slope Intercept form is $y \geq -\frac{1}{2}x + 3$, so the line is a **solid line** and the ≥ requires that the **area above the line** is shaded. The y-intercept is the point (0, 3) and the slope m is $-\frac{1}{2}$.

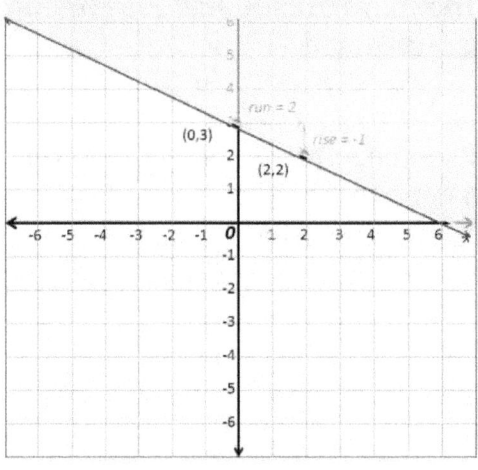

Figure A.7.1

 INTEGRATING TECHNOLOGY    CULTURAL COMPETENCY

 PROJECT-BASED LEARNING    WRITING ACROSS THE CURRICULUM

www.keepingitrealeasymatch.com

KEEPING IT REAL - ALGEBRA I   A.7(C)

Student Materials

Objectives for A.7 (C): Determine Reasonableness of Solutions to Linear Inequalities

1. The highest temperature recorded in the U.S. was 130 degrees Fahrenheit. Use this data to graph the past temperatures in the U.S.

2. The average annual snow fall for a city is 35 inches. The first 10 weeks of winter, the city received a total of 12 inches of snow. It is expected to snow and average of 2.5 inches per week through the end of the winter. What is the reasonable number of weeks needed for the city to reach its average annual snow fall?

_____
_____
_____
_____

KEEPING IT REAL - ALGEBRA I   A.7(C)

3. Solve the following inequalities, graph and check your solutions:

   a. $x - 5 > -3$

   b. $x - 8 \leq 2$

   c. $x - 9 < -4$

   d. $3x - 7 < 6$

   e. $y > \dfrac{1}{4}x + 2$

   f. $y > 3x - 2$

   g. $y < x - 3$

   h. $y \leq -\dfrac{1}{4}x - 3$

   i. $y \leq -x - 3$

EasyMatch A.8 (A)

**STAAR – Algebra I: Linear Equations and Inequalities.
The student will demonstrate an understanding of linear functions.**

A.8–Linear functions. The student formulates systems of linear equations from problem situations, uses a variety of methods to solve them, and analyzes the solutions in terms of the situation. The student is expected to

A.8 (A) - Analyze situations and formulate systems of linear equations in two unknowns to solve problems; **Supporting Standard**

STANDARD AND LESSON OVERVIEW
TEACHER DIRECTIONS

A.8 (A) - Analyze situations and formulate systems of linear equations in two unknowns to solve problems; **Supporting Standard**

A system of linear equations, or linear system, consists of two or more linear equations in the same variables. **A solution of a linear system is an ordered pair that satisfies each equation in the system.** One method of solving a system of linear equations is by finding **a Point of Intersection – the solution. A system of linear equations has no solution when the graphs of the equations are parallel – have the same slope. There is no point of intersection, so there is no solution.** The system of equations can be formulated from a given situation.

For example, the YMCA offers a season pass for $100.00. There are two situations associated with the season pass.

a. The pass holder pays $5 per access to all facilities.
b. Without the season pass, a person pays $10 dollars per access to all facilities.

A system of equations can determine the number x of accesses to all facilities for which the total cost y with the season pass, including the cost of the pass, is the same as the total cost without the season pass.

KEEPING IT REAL - ALGEBRA I A.8(A)

The solution involves writing a system of equations where y is the total cost for x sessions.

Equation One: $y = 10x$ Accesses without the season pass

Equation Two: $y = 100 + 5x$ Accesses with the season pass

 INTEGRATING TECHNOLOGY    CULTURAL COMPETENCY

 PROJECT-BASED LEARNING    WRITING ACROSS THE CURRICULUM

www.keepingitrealeasymatch.com

KEEPING IT REAL - ALGEBRA I

A.8(A)

Student Materials

Objectives for A.8 (A): Formulate a System of Linear Equations

Formulate a system of equations for each situation below.

1. A business rents skates and bicycles. On one day, the business has a total of 35 rentals and collects $500 for rentals. Find the number of pairs of skates rented and the number of bicycles rented if the skates are $10 per day and the bicycles are $25 per day.

_____
_____
_____
_____
_____
_____

2. There are 10 books in a box. The thickness of each book is either 1 inch or 2 inches. The height of the box is 16 inches. Write a system of equations to determine x, the number of 1 inch-thick books, and y, the number of 2 inch-thick books.

_____
_____
_____
_____
_____
_____

KEEPING IT REAL - ALGEBRA I   A.8(A)

3. The ratio of boys to girls in a 9th grade class at a middle school is 3 to 2. There are 600 9th graders. How many boys and how many girls?

_____

_____

_____

_____

_____

EasyMatch A.8 (B)

**STAAR – Algebra I: Linear Equations and Inequalities.
The student will demonstrate an understanding of linear functions.**

A.8 – Linear functions. The student formulates systems of linear equations from problem situations, uses a variety of methods to solve them, and analyzes the solutions in terms of the situation. The student is expected to

A.8 (B) - Solve systems of linear equations using [concrete] models, graphs, tables, and algebraic methods; **Readiness Standard**

STANDARD AND LESSON OVERVIEW
TEACHER DIRECTIONS

A.8 (B) - Solve systems of linear equations using [concrete] models, graphs, tables, and algebraic methods; **Readiness Standard**

System of Linear Equations - Point of Intersection

A system of linear equations consists of two or more linear equations with the same variables. One method of solving a system of linear equations is by finding **a Point of Intersection – the solution. A system of linear equations has no solution when the graphs of the equations are parallel. There is no point of intersection, so there is no solution.** Consider the two equations below:

$2x - 3y = 4$ Equation One

$x + 2y = -5$ Equation Two

Notes

$2x - 3y = 4$ Solve by elimination using the Additive Inverse

$-2(x + 2y = -5)$ Multiply Equation Two by a -2 and eliminate the x variable.

$2x - 3y = 4$
$-2x - 4y = 10$ Additive Inverse: $2x - 2x$
$0 - 7y = 14$ Spirals back to Multiplicative Inverse and Distributive Property

KEEPING IT REAL - ALGEBRA I A.8(B)

$-\frac{1}{7}(-7y = 14)$ Multiplicative Inverse: multiply by $-\frac{1}{7}$ and Distributive Property

y = - 2   **Solved for y**

The next step is to substitute -2 back into either equation for y and solve for x.

2x – 3(-2) = 4   Substituting -2 for y in Equation One
2x + 6 = 4
2x + 6 – 6 = 4 – 6   Additive Inverse; add -6 to both sides

$\frac{1}{2}(2x = -2)$ Multiplicative Inverse; multiply by $\frac{1}{2}$, and Distributive Property

x = -1   **The solution is (-1, -2) and is the Point of Intersection**.

System of Linear Equations - Elimination

2x – 3y = 4   **Equation One**
x + 2y = -5   **Equation Two**

Notes

Consider equations one and two above. Eliminate the y variable using the Commutative Property of real numbers, which states that if a and b are real numbers, then a x b = b x a.

2x – 3y = 4   To eliminate y, multiply Equation One by 2 and
x + 2y = -5   multiply Equation Two by 3

2(2x – 3y = 4)   Apply the Distributive Property
3(x + 2y = -5)

4x – 6y = 8   - 6y + 6y = 0
3x + 6y = -15)

7x = -7   Apply Multiplicative Inverse and
$\frac{1}{7}(7x = -7)$ Distributive Property

x = -1

-1 + 2y = -5   Substitute -1 for x in Equation Two

KEEPING IT REAL - ALGEBRA I A.8(B)

+1 -1 + 2y = -5 +1	Additive Inverse, add 1 to both sides
2y = -4	Apply the Multiplicative Inverse and then the
$\frac{1}{2}(2y = -4)$	Distributive Property
y = -2	Solved for y, the **solution is (-1, -2) the point of intersection**

System of Linear Equations - Slope Intercept Form: y = mx + b

2x - 3y = 4	**Equation One**
x + 2y = -5	**Equation Two**

Notes

Transforming equations to the slope intercept form: **y = mx + b**

-3y = -2x + 4	Equation One first, Additive inverse, -2x to both sides then apply
$\frac{-1}{3}(-3y = -2x + 4)$	the Multiplicative Inverse and Distributive Property
$y = \frac{2}{3}x - \frac{4}{3}$	**Equation One in Slope Intercept Form**
x + 2y = -5	Transforming Equation Two
2y = -x - 5	Additive Inverse, -x to both sides and then apply the
$\frac{1}{2}(2y = -x - 5)$	Multiplicative Inverse and Distributive Property
$y = -\frac{1}{2}x - \frac{5}{2}$	**Equation Two in Slope Intercept Form**
y = y	**Set y = y with respect to Equation One and Equation Two**
$\frac{2}{3}x - \frac{4}{3} = -\frac{1}{2}x - \frac{5}{2}$	First, combine like terms using the Additive Inverse
$\frac{2}{3}x + \frac{1}{2}x = \frac{4}{3} - \frac{5}{2}$	Combined like terms (Additive Inverse), find a common denominator
$\frac{4+3}{6}x = \frac{8-15}{6}$	Properties of fractions, $\frac{a}{b} + \frac{c}{d} = \frac{ad + cb}{bd}$ or $\frac{a}{b} - \frac{c}{d} = \frac{ad - cb}{bd}$

KEEPING IT REAL - ALGEBRA I A.8(B)

$\frac{7}{6}x = \frac{-7}{6}$ Combining numerator terms, after finding a common denominator

$\frac{6}{7}(\frac{7}{6}x = \frac{-7}{6})$ Multiplicative Inverse and Distributive Property

x = -1 **Solved for x**

Notes

$y = -\frac{1}{2}x - \frac{5}{2}$ With x = -1, y (-1) follows: Equation Two in Slope Intercept form

$y(-1) = -\frac{1}{2}(-1) - \frac{5}{2}$ Substituting -1 for x

$y(-1) = \frac{1}{2} - \frac{5}{2}$ Property of fractions

$y(-1) = -\frac{4}{2}$ Common denominator, so add the numerators

y(-1) = - 2 **The solution is (-1, -2) the point of intersection**

Use a Graph to find the Solution to a System of Equations
Graph the Slope Intercept forms of Equation One and Equation Two from above

$y = \frac{2}{3}x - \frac{4}{3}$ **Equation One in Slope Intercept Form**

$y = -\frac{1}{2}x - \frac{5}{2}$ **Equation Two in Slope Intercept Form**

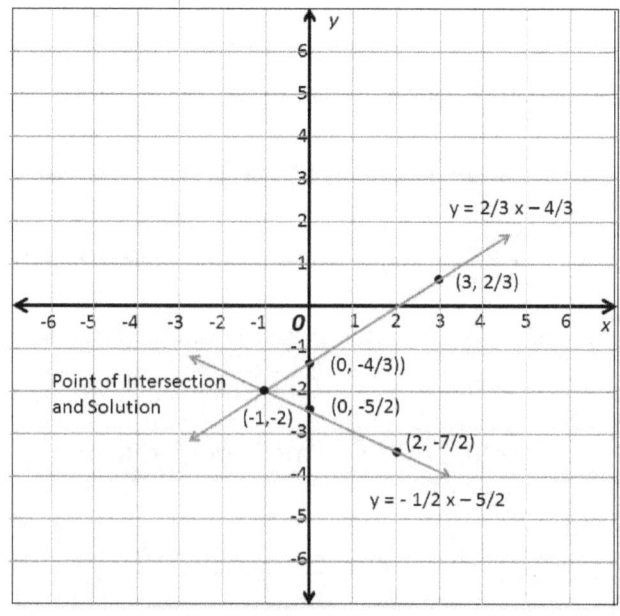

Figure A.8.1

KEEPING IT REAL - ALGEBRA I   A.8(B)

Figure A.8.1 illustrates the lines for Equation One and Equation Two. The Point of Intersection P(-1, -2) is the solution for the two equations, when solving as a System of Equations, or a Linear System.

Solving a System of Equations Using the Table Method

Another method to solve a system of linear equations is to make a table of points representing each line and then evaluate the points to find the same point for both lines:

$2x - 3y = 4$ **Equation One**
$x + 2y = -5$ **Equation Two**

$y = \dfrac{2}{3}x - \dfrac{4}{3}$ **Equation One in Slope Intercept Form**

$y = -\dfrac{1}{2}x - \dfrac{5}{2}$ **Equation Two in Slope Intercept Form**

One could use the TI-84 Plus calculator and trace the graph of each equation. The following points will be identified as the lines are traced. As shown in Table A.8.1, the two lines share the same point of (-1, -2) which is the point of intersection and also the solution for the linear system of equations. Another method is to algebraically solve for y(x) for each equation.

Equations			
One		Two	
x	y	x	y
0	1.3	0	-2.5
-1	-2	-1	-2
-2	-2.6	-7	-1
2	0	-5	0
4	1.3	-3	-4

Table A.8.1

 INTEGRATING TECHNOLOGY    CULTURAL COMPETENCY

 PROJECT-BASED LEARNING    WRITING ACROSS THE CURRICULUM

KEEPING IT REAL - ALGEBRA I   A.8(B)

Student Materials

Objectives for A.8 (B): Solving a System of Linear Equations

1. Solve the following systems of linear equations using the y-intercept model, graphs, and tables.

    a. $2x - 2y = 4$ $3x + 2y = -3$

    b. $y = 2x - 3$ $y = x - 1$

    c. $3x - y = 17$ $2x + y = 8$

    d. $x + 3y = 5$ $2x - 4y = -5$

    e. $y = -2x - 3$ $y = x$

2. Solve for the following systems of linear equations using algebraic methods: Elimination and Point of Intersection.

    f. $2x - 2y = 4$ $3x + 2y = -3$

    g. $y = 2x - 3$ $y = x - 1$

    h. $3x - y = 17$ $2x + y = 8$

    i. $x + 3y = 5$ $2x - 4y = -5$

    j. $y = -2x - 3$ $y = x$

KEEPING IT REAL - ALGEBRA I A.8(B)

3. Draw a graph of two lines representing a system of equations that has no solution.

4. For the system of equations below, determine the value of x.

$$12x + 6y = 26$$
$$6x - 2y = 8$$

5. The cheerleaders conducted a bake sale for practice uniforms. The number of cakes sold was seven more than twice the number of pies sold. The cheerleader sold a total of 55 cakes and pies. How many pies were sold?

_____
_____
_____
_____

KEEPING IT REAL - ALGEBRA I   A.8(B)

EasyMatch A.8 (C)

**STAAR – Algebra I: Linear Equations and Inequalities.
The student will demonstrate an understanding of linear functions.**

A.8 – Linear functions. The student formulates systems of linear equations from problem situations, uses a variety of methods to solve them, and analyzes the solutions in terms of the situation. The student is expected to

A.8 (C) - Interpret and determine the reasonableness of solutions to systems of linear equations. **Supporting Standard**

STANDARD AND LESSON OVERVIEW
TEACHER DIRECTIONS

A.8 (C) - Interpret and determine the reasonableness of solutions to systems of linear equations. **Supporting Standard**

Parallel Lines - symbol (∥)

A key concept states that if two non-vertical lines in the same plane have the same slope, then the lines are parallel. If two non-vertical lines in the same plane are parallel, then the lines have the same slope. **A system of linear equations has no solution when the graphs of the equations are parallel. There is no point of intersection, so there is no solution.**

For example, write the equation of the line that passes through the point (1, -2) and is **parallel** to the line $y = 3x - 2$.

$y = 3x - 2$ and Point (1, -2)	Notes
$y = 3x - 2$	Slope is 3
$y = mx + b$	Slope intercept form
$-2 = 3(1) + b$	Substitute 3 for m, 1 for x and -2 for y
$b = -5$	Solving for b
$y = 3x - 5$	Using $y = mx + b$, substitute 3 for m and -5 for b

KEEPING IT REAL - ALGEBRA I   A.8(C)

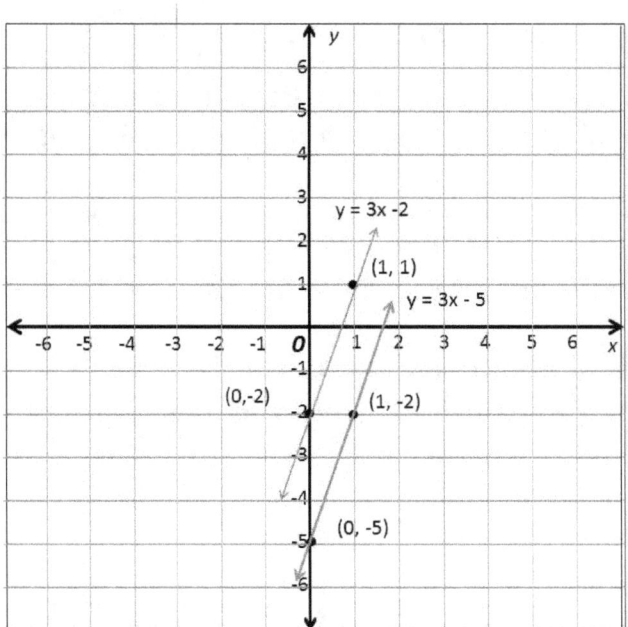

Figure A.8.2

The reasonable solution from examining the graph is for the line of equation **y = 3x -2**, the line which is parallel, having the same slope of 3, and passes through point (1, -2) is the line for equation **y = 3x -5**.

Perpendicular Lines – symbol (⊥)
Another key concept involves lines which are perpendicular. Two lines in the same plane that intersect to form a right (90 degrees) are perpendicular. For two non-vertical lines in the same plane which have slopes that are **negative reciprocals**, then the lines are perpendicular. If the two non-vertical lines in the same plane are perpendicular, then their slopes are **negative reciprocals**. For example, write the equation of the line that passes through the point (1, -2) and is **perpendicular** to the line y = 3x -2.

y = 3x -2 and Point (1, -2)   Notes

$y = 3x - 2$ Slope is 3. Thus $-\dfrac{1}{3}$, the negative reciprocal of 3, is slope for ⊥ line

$y = mx + b$ Using the slope intercept form, to solve for b

$-2 = -\dfrac{1}{3}(1) + b$ Substitute $-\dfrac{1}{3}$ for m and from the point (1, -2), 1 for x and -2 for y

$b = -\dfrac{5}{3}$ Solving for b, Additive Inverse by adding $\dfrac{1}{3}$ to both sides

KEEPING IT REAL - ALGEBRA I   A.8(C)

$y - y_1 = m(x - x_1)$ Alternative, use the equation for the slope of a line to solve for b

$y - (-2) = -\dfrac{1}{3}(x - 1)$ Substituting the values of the slope and of point (1, -2) for x_1 and y_1

$y + 2 = -\dfrac{1}{3}x + \dfrac{1}{3}$ Distributive property and Additive Inverse, -2 added to both sides

$y = -\dfrac{1}{3}x - \dfrac{5}{3}$ Using the y = mx + b form, $b = -\dfrac{5}{3}$, see Figure A.8.3 below.

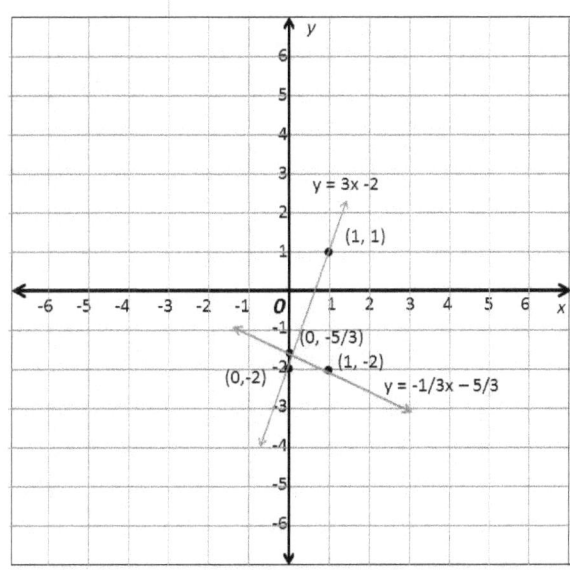

Figure A.8.3

Solving a System of Equations Using the Substitution Method (y = y)
In solving a system of linear equations, another method to use is substitution. The substitution method, allows one to develop a one-variable equation. The approach is to take one of the equations in the y-intercept form and substitute for y in the other equation.

For example, consider the following system of equations:

System	Notes
y = x + 6.1	Equation One
y = -2x - 1.4	Equation Two
y = -2x - 1.4	Start with one of the equations, Equation Two
x + 6.1 = -2x -1.4	Substitute the expression for y, x + 6.1, from One into Equation Two
x + 6.1 = -2x -1.4	Resulting in a one-variable equation, solve for x
3x = -7.5	Additive Inverse, add 2x to both sides and subtract 6.1 from both sides

KEEPING IT REAL - ALGEBRA I   A.8(C)

x = -2.5   **Solved for x,** substitute -2.5 for x in either equation and solve for y
y = -2.5 + 6.1   **Solving for y, using Equation One**
y = 3.6   Solved for y, the solution is (-2.5, 3.6)

 INTEGRATING TECHNOLOGY

 CULTURAL COMPETENCY

 PROJECT-BASED LEARNING

 WRITING ACROSS THE CURRICULUM

www.keepingitrealeasymatch.com

KEEPING IT REAL - ALGEBRA I   A.8(C)

Student Materials

Objectives for A.8 (C): Determine the Reasonableness of Solutions of Linear Systems

1. Given the equation $2x + 3y = 6$ and a Point $(3, -4)$, write the equation of a line which is parallel and the equation of a line which is perpendicular to the given equation and passing through the given point.

___

2. Given the equation $2x - 4y = 8$ and a Point $(3, -2)$, write the equation of a line which is parallel and the equation of a line which is perpendicular to the given equation and passing through the given point.

___

3. Solve the system of equations using substitution and determine the reasonableness of the solutions.

    a. $y = 2x$ $6x - y = 8$

    b. $y = 3x + 1$ $x = 3y + 1$

    c. $2y = 0.2x + 7$ $3y - 2x = 2$

KEEPING IT REAL - ALGEBRA I A.8(C)

4. The high school graduating class needs a combined total of 432 medium and large T-shirts. The number of medium T-shirts needs to be three times the number of large T-shirts. Would it be correct for graduating class to order 144 large T-shirts and 288 medium T-shirts? Explain your answer.

_____
_____
_____
_____
_____
_____
_____

> # EasyMatch A.9 (A)
>
> **STAAR – ALGEBRA 1: Quadratic and other nonlinear functions.**
> **The student will demonstrate an understanding of quadratic and other non-linear functions.**
>
> **A.9 - Quadratic and other nonlinear functions.** The student understands that the graphs of quadratic functions are affected by the parameters of the function and can interpret and describe the effects of changes in the parameters of quadratic functions. The student is expected to
>
> A.9(A) - Determine the domain and range for quadratic functions in given situations; **Supporting Standard**

STANDARD AND LESSON OVERVIEW
TEACHER DIRECTIONS

A.9(A) - Determine the domain and range for quadratic functions in given situations; **Supporting Standard**

Quadratic Functions and Equations

A quadratic function is a nonlinear function that can be written in the standard form $y = ax^2 + bx + c$, where $a \neq 0$. Every quadratic function has a U-shaped graph called a parabola. The **parent quadratic function is $y = x^2$**. If the coefficient of the x^2 term is positive, the graph will turn up at the **vertex** - the lowest or highest point of the parabola. The vertex for the graph of $y = x^2$ is point (0, 0).

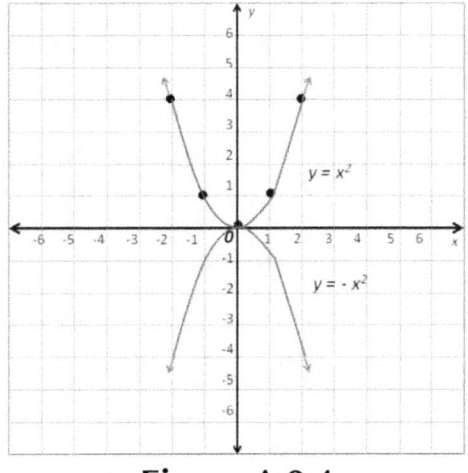

Figure A.9.1

KEEPING IT REAL - ALGEBRA I   A.9(A)

The **domain** of the function $y = x^2$ is all Real Numbers. The **range** is all $y \geq 0$. If $y = -x^2$, the graph turns down and the domain is all Real Numbers and the range is all $y \leq 0$.
If we consider the graph of $y = x^2 + 4$, the graph of $y = x^2$ is shifted up four (4) units.

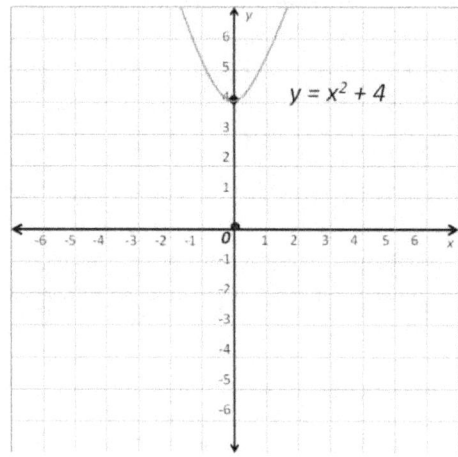

Figure A.9.2

The **domain** is all real numbers and the **range** is all real numbers for $y \geq 4$.

The line that passes through the vertex and divides the parabola into two symmetric parts is called the **axis of symmetry**. The **axis of symmetry** for the graph $y = x^2$ is the y-axis, where $x = 0$.

Absolute Value

The absolute value of a number **a**, written |**a**|, is the distance between **a** and 0. An absolute value equation is an equation that contains an absolute value expression such as, $|x| = 2$. The equation $|x| = 2$ means that the distance between **x** and 0 is 2. The key concepts are:

    If **a** is positive, then the |**a**| = **a**.   Ex. |3| = 3
    If **a** is 0, then |**a**| = 0.   Ex. |0| = 0
    If **a** is negative, then |**a**| = -**a**.   Ex. |-2| = - (-2) = 2

In solving an absolute value equation, $|ax + b| = c$, where $c \geq 0$, is equivalent to the statement $ax + b = c$ or $ax + b = -c$. For example:

Solve for $\|x - 3\| = 8$	**Notes**
$\|x - 3\| = 8$	Write the original equation
$x - 3 = 8$ or $x - 3 = -8$	Rewrite as two equations
$x = 11$ or $x = -5$	Additive Inverse, add 3 to each side
$x = 11$ or $x = -5$	The solutions are 11 and -5

KEEPING IT REAL - ALGEBRA I A.9(A)

$\|11 - 3\| = 8$ or $\|-5 - 3\| = 8$	Checking the solutions
$\|8\| = 8$ or $\|-8\| = 8$	Subtract, respectively
8 = 8 or 8 = 8	**The solutions check**

Zero-Product Property

The Zero-Product Property is a key concept in solving polynomial equations in factored form. Let *a* and *b* be real numbers. If *ab* = 0, then *a* = 0 or *b* = 0. The Zero-Product Property is used to solve an equation when one side is zero and the other side is a product of polynomial factors. **The solutions of such an equation are also called the roots. Factoring** is used to solve a polynomial equation using the zero-product property. **To factor the polynomial, write it as a product of other polynomials.** Look for the greatest common factor (GFC) of the polynomial's terms. This is the monomial with an integer coefficient that divides evenly into each term.

Given the quadratic equation: $x^2 - 2x - 3 = 0$

Notes

$x^2 - 2x - 3 = 0$	$y = ax^2 + bx + c$, is the standard form
$(x-3)(x+1) = 0$	**Factoring, writing as product of other polynomials**
$x-3=0 \quad x+1=0$	Using the zero-product property, set both results equal to 0
$x = 3 \quad x = -1$	Thus, the graph will cross the x axis at points (3, 0) and (-1, 0)
$x = \dfrac{-(-2)}{2(1)}$	To find the vertex, the x component of the vertex is: $x = \dfrac{-b}{2a}$
$x = 1$	Using $f(x) = x^2 - 2x - 3$, to find the y component of the vertex
$f(1) = (1)^2 -2(1) - 3$	by substituting the value of x, to find f of x, or f of (x = 1)
$f(1) = 1 - 2 - 3$	
$f(1) = -4$	The y component is -4, thus the Vertex is (1, -4)

The graph of the equation $x^2 - 2x - 3 = 0$, is illustrated in Figure A.9.3. The **domain** is all real numbers and the **range** is all y ≥ -4.

KEEPING IT REAL - ALGEBRA I   A.9(A)

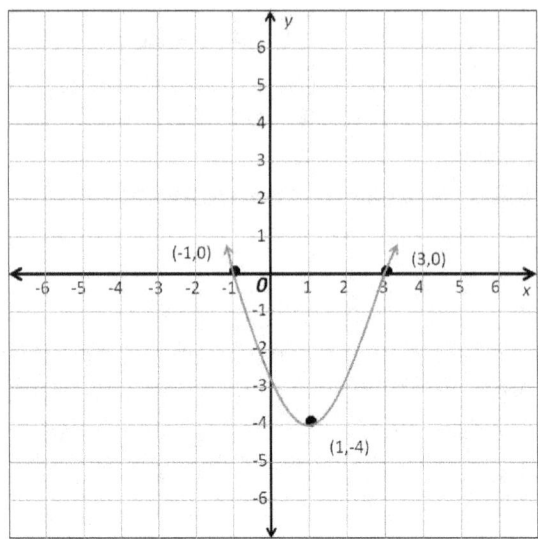

Figure A.9.3

Completing the Square

Given the general form of a quadratic equation **ax² + bx + c = 0 (where a ≠ 0)**, the procedure for completing the square is presented below.

ax² + bx + c = 0, (where a ≠ 0)   Notes

$ax^2 + bx = -c$	Additive Inverse, subtracting **c** from both sides
$\frac{1}{a}(ax^2 + bx = -c)$	Multiplicative Inverse, to make the coefficient of **x²** one (1)
$x^2 + \frac{b}{a}x = -\frac{c}{a}$	and the Distributive Property, results in this equation
$x^2 + \frac{b}{a}x = -\frac{c}{a}$	Completing the Square: Take half ($\frac{1}{2}$) of the coefficient of x, square that value and add to both sides
$x^2 + \frac{b}{a}x = -\frac{c}{a}$	Half of $\frac{b}{a}$ is $\frac{b}{2a}$, squaring results in $(\frac{b}{2a})^2 = \frac{b^2}{4a^2}$
$x^2 + \frac{b}{a}x + \frac{b^2}{4a^2} = \frac{b^2}{4a^2} - \frac{c}{a}$	Adding, $\frac{b^2}{4a^2}$, to both sides
$x^2 + \frac{b}{a}x + \frac{b^2}{4a^2} = \frac{b^2 - 4ac}{4a^2}$	Using 4a² as the common denominator, $-\frac{4ac}{4a^2} = -\frac{c}{a}$

KEEPING IT REAL - ALGEBRA I A.9(A)

$(x+\frac{b}{2a})^2 = \frac{b^2-4ac}{4a^2}$ In completing the square, the left side of the equation a perfect trinomial square.

$\sqrt{(x+\frac{b}{2a})^2} = \pm\sqrt{\frac{b^2-4ac}{4a^2}}$ Take the square root of both sides, indicates a ± condition for the right side of the equation

$x+\frac{b}{2a} = \pm\frac{\sqrt{b^2-4ac}}{2a}$ Take the square root of the left side and the denominator

$x = -\frac{b}{2a} \pm \frac{\sqrt{b^2-4ac}}{2a}$ Additive Inverse, subtracting $\frac{b}{2a}$ from both sides results in the

$x = \frac{-b \pm \sqrt{b^2-4ac}}{2a}$ Quadratic Formula, with $x=\frac{-b}{2a}$ for the x component of the vertex and evaluating f $(\frac{-b}{2a})$, gives the y value of the vertex.

Another Approach

For a quadratic expression of the form x² + bx, one can add a constant c to the expression so that the expression x² + bx + c is a perfect trinomial square. The process is called completing the square. To complete the square for an expression x² + bx, add the square of half the coefficient the bx term. **Note: Check that the coefficient of the x² term is one (1).**

For example, $x^2 + bx + (\frac{b}{2})^2 = (x+\frac{b}{2})^2$

In finding the value of c that makes the expression **x² + 5x + c** a perfect trinomial square use the procedure above.

                                    Notes
x² + 5x + c   First find the value for **c**

$c = (\frac{5}{2})^2 = (\frac{25}{4})$ Find the square of half of the coefficient of x, or 5

$x^2+5x+c = x^2+5x+\frac{25}{4}$ Substitute $\frac{25}{4}$ for **c**

$= (x+\frac{5}{2})^2$ **Square of a binomial**

KEEPING IT REAL - ALGEBRA I A.9(A)

 INTEGRATING TECHNOLOGY    CULTURAL COMPETENCY

 PROJECT-BASED LEARNING    WRITING ACROSS THE CURRICULUM

www.keepingitrealeasymatch.com

KEEPING IT REAL - ALGEBRA I   A.9(A)

Student Materials

Objectives for A.9 (A): Determine the Domain and Range of the Quadratic Functions

1. Graph the following quadratic functions and determine the domain and the range.

   a. $y = x^2 - 4$ b. $y = x^2 - 2$ c. $y = x^2 + 2$

   d. $y = x^2 + 9$ e. $y = -\dfrac{1}{5}x^2$ f. $y = -\dfrac{3}{2}x^2$

2. Find the vertex and axis of symmetry.

   a. $y = 3x^2 - 6x + 2$

   b. $y = x^2 + 2x + 8$

KEEPING IT REAL - ALGEBRA I   A.9(A)

3. Find the value of **c** that makes the expression a perfect trinomial square. Then write the expression as a square of a binomial.

    a. $x^2 + 12x + c$ b. $x^2 - 9x + c$ c. $x^2 - 18 + c$

EasyMatch A.9 (B)
STAAR – Algebra I: Quadratic and Other Nonlinear Functions
The student will demonstrate an understanding of quadratic and other nonlinear functions.

A.9 - Quadratic and other nonlinear functions. The student understands that the graphs of quadratic functions are affected by the parameters of the function and can interpret and describe the effects of changes in the parameters of quadratic functions. The student is expected to

A.9 (B) - Investigate, describe, and predict the effects of changes in a on the graph of $y = ax^2 + c$; **Supporting Standard**

STANDARD AND LESSON OVERVIEW
TEACHER DIRECTIONS

A.9 (B) - Investigate, describe, and predict the effects of changes in a on the graph of $y = ax^2 + c$; **Supporting Standard**

The function $y = ax^2 + c$, is used to illustrate the changes with the graph for a quadratic function. With **a** and **c** as parameters of the function, various changes will have certain impacts.

Note the comparison of the graphs for the two functions below.

$y = \frac{1}{2}x^2 - 4$ and $y = x^2$ See Figure A.9.4.

Make a table of values for $y = \frac{1}{2}x^2 - 4$

x	-4	-2	0	2	4
y	4	-2	-4	-2	4

KEEPING IT REAL - ALGEBRA I   A.9(B)

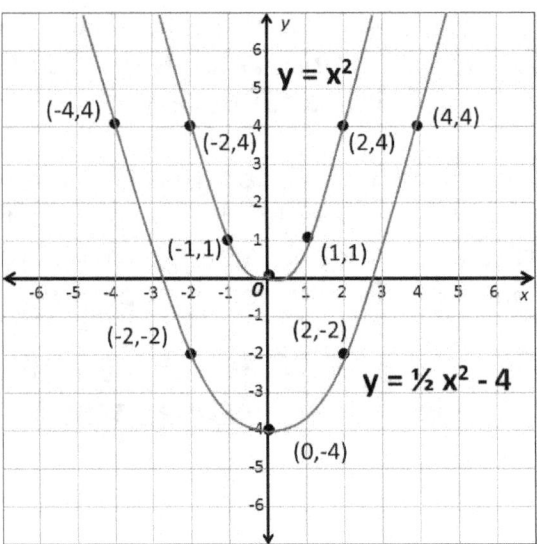

Figure A.9.4

Notes:
1. The smaller the coefficient of x^2, the **a** parameter, the wider the graph opens.
2. The coefficient of x^2, the **a** parameter, determines the rate in which the domain increases.
3. The **c** parameter moves the vertex of the graph up if positive or down if negative.
4. The **c** parameter impacts the range of the function.

www.keepingitrealeasymatch.com

KEEPING IT REAL - ALGEBRA I   A.9(B)

Student Materials

Objectives for A.9 (B): Investigate the changes of a in the graph $y = ax^2 + c$.

1. Graph the following functions and compare with the graph of $y = x^2$.

Use the values for x as provided in the table below for each quadratic function

x	-2	-1	0	1	2
y					

a. $y = 2x^2 - 4$

b. $y = -2x^2 + 1$

c. $y = \dfrac{1}{4}x^2 - 2$

d. $y = \dfrac{1}{2}x^2 - 4$

KEEPING IT REAL - ALGEBRA I   A.9(B)

2. Graph the function $y = 3x^2 - 2$. How would the graph change, if the coefficient of x^2 were changed from 3 to another positive number?

_____
_____
_____
_____

3. Graph examples for the following quadratic expressions and discuss the impact of the change in the *a* parameter.

a. $y = ax^2$ where $a > 1$, $a = 1$ and $0 < a < 1$

b. $y = ax^2$ where $a < -1$, $a = -1$ and $-1 < a < 0$

_____
_____
_____
_____

EasyMatch A.9 (C)

STAAR – Algebra I: Quadratic and Other Nonlinear Functions
The student will demonstrate an understanding of quadratic and other nonlinear functions.

A.9 - Quadratic and other nonlinear functions. The student understands that the graphs of quadratic functions are affected by the parameters of the function and can interpret and describe the effects of changes in the parameters of quadratic functions. The student is expected to

A.9 (C) - Investigate, describe, and predict the effects of changes in c on the graph of $y = ax^2 + c$; **Supporting Standard**

STANDARD AND LESSON OVERVIEW
TEACHER DIRECTIONS

A.9 (C) - Investigate, describe, and predict the effects of changes in c on the graph of $y = ax^2 + c$; **Supporting Standard**

As presented in A.9(B), the function $y = ax^2 + c$, is used to illustrate the changes with the graph for a quadratic function. With **a** and **c** as parameters of the function, various changes will have certain impacts.

The c parameter effects changes in the $y = ax^2 + c$, with **an upward vertical translation if c > 0 and a downward vertical translation in c < 0.**

Note the comparison of the graphs for the two functions below.

$y = \dfrac{1}{2}x^2 - 4$ and $y = x^2$ See Figure A.9.4.

Make a table of values for $y = \dfrac{1}{2}x^2 - 4$

x	-4	-2	0	2	4
y	4	-2	-4	-2	4

KEEPING IT REAL - ALGEBRA I A.9(C)

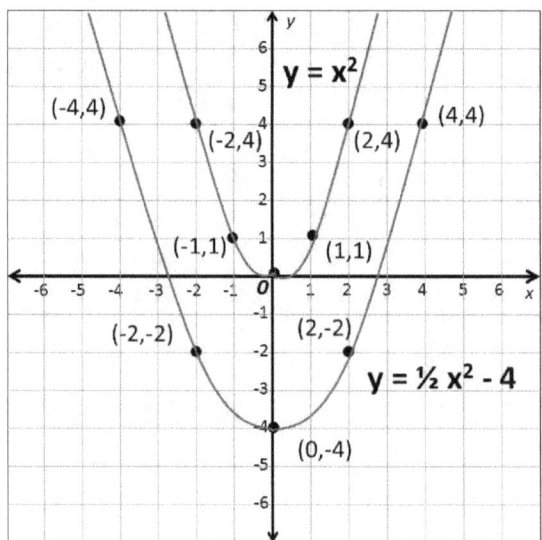

Figure A.9.4

Notes:
1. The smaller the coefficient of x^2, the **a** parameter, the wider the graph opens.
2. The coefficient of x^2, the **a** parameter, determines the rate in which the domain increases.
3. The **c** parameter moves the vertex of the graph **up if positive** or **down if negative**.
4. The **c** parameter impacts the range of the function.

 INTEGRATING TECHNOLOGY    CULTURAL COMPETENCY

 PROJECT-BASED LEARNING    WRITING ACROSS THE CURRICULUM

www.keepingitrealeasymatch.com

KEEPING IT REAL - ALGEBRA I A.9(C)

Student Materials

Objectives for A.9 (C): Investigate the changes of c in the graph $y = ax^2 + c$.

1. Graph the following functions and compare with the graph of **y = x²**.

Use the values for x as provided in the table below for each quadratic function

x	-2	-1	0	1	2
y					

a. $y = 2x^2 - 5$

b. $y = -2x^2 + 2$

c. $y = \dfrac{1}{4}x^2 - 1$

d. $y = \dfrac{1}{2}x^2 - 2$

KEEPING IT REAL - ALGEBRA I　　　　A.9(C)

2. Graph the function $y = 3x^2 - 4$, using the values in the table in question 1. How would the graph change, if the **c** parameter were changed from -4 to another negative number?

_____
_____
_____
_____

3. Graph examples for the following quadratic expressions and discuss the impact of the change in the **c** parameter.

   $y = x^2 + c$, where $c > 0$, $c = 0$ and $c < 0$

_____
_____
_____
_____

KEEPING IT REAL - ALGEBRA I   A.9(C)

4. Graph the functions below and compare the graph with the graph of $y = x^2$.

   a. $y = x^2 - 6$

   b. $y = x^2 + 5$

   c. $y = x^2 - 2$

EasyMatch A.9 (D)

STAAR – Algebra I: Quadratic and Other Nonlinear Functions The student will demonstrate an understanding of quadratic and other nonlinear functions.

A.9 - Quadratic and other nonlinear functions. The student understands that the graphs of quadratic functions are affected by the parameters of the function and can interpret and describe the effects of changes in the parameters of quadratic functions. The student is expected to

A.9 (D) - Analyze graphs of quadratic functions and draw conclusions.
Readiness Standard

STANDARD AND LESSON OVERVIEW
TEACHER DIRECTIONS

A.9 (D) - Analyze graphs of quadratic functions and draw conclusions. **Readiness Standard**

When analyzing the graphs of quadratic functions, the graphs are often compared to the **parent quadratic function $y = x^2$**. The quadratic function $y = ax^2 + c$ was presented to analyze the effects of change in parameters **a** and **c**. The graphs presented in A.9.1-4 are provided to analyze and draw various conclusions.

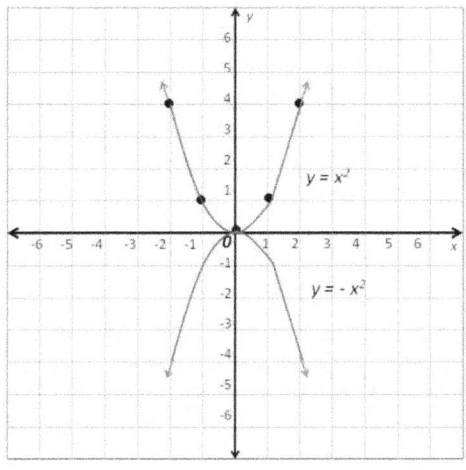

Figure A.9.1

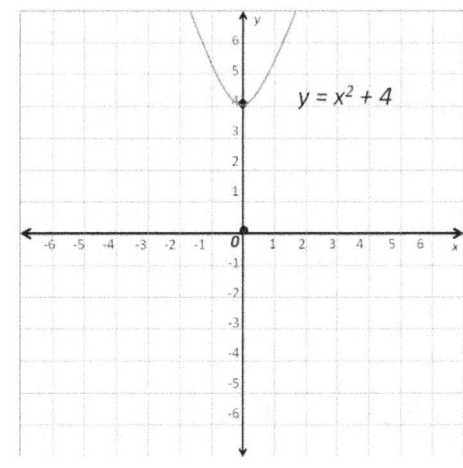

Figure A.9.2

KEEPING IT REAL - ALGEBRA I — A.9(D)

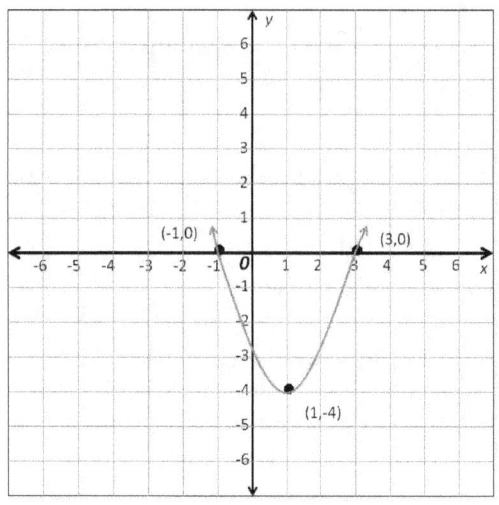

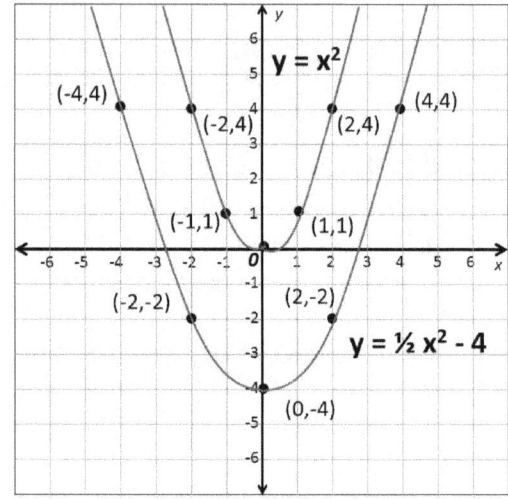

Figure A.9.3   **Figure A.9.4**

Some Conclusions (not all conclusive):

Figure A.9.1: $y = x^2$ has vertex at point (0, 0) and turns up for positive **a**, coefficient of x^2.

Figure A.9.2: $y = x^2 + c$, the **c** parameter effects a vertical translation of the graph (up/down)

Figure A.9.3: Factoring and the zero-product property are used to find the x components and $x = \dfrac{-b}{2a}$ to find the x component of the vertex, $f(x) = y$ component

Figure A.9.4: $y = ax^2 + c$, the graph widens for a small value of the **a** parameter

 INTEGRATING TECHNOLOGY    CULTURAL COMPETENCY

 PROJECT-BASED LEARNING    WRITING ACROSS THE CURRICULUM

www.keepingitrealeasymatch.com

KEEPING IT REAL - ALGEBRA I   A.9(D)

Student Materials

Objectives for A.9 (D): Analyze graphs of quadratic functions and draw conclusions.

1. Graph the following functions below and compare the graph with the graph of $y = x^2$ function.

   a. $y = 6x^2 + 5$

   b. $y = x^2 + 3$

   c. $y = -3x^2 - 2$

   d. $y = x^2 + 1.75$

   e. $y = \dfrac{1}{2}x^2 - 4$

2. Solve the following equations by graphing. Make a table for the x and y values.

   a. $-x^2 - 6x = 0$
   b. $2x^2 = 2$
   c. $x^2 - 7x + 10 = 0$

KEEPING IT REAL - ALGEBRA I A.9(D)

3. Tell whether the graph opens upward or downward. Then find the axis of symmetry and the vertex of the graph of the function.

    a. $y = x^2 - -5$

    b. $y = -x^2 + 9$

    c. $y = 3x^2 + 6x - 2$

    d. $y = -2x^2 + 7x - 21$

    e. $y = \dfrac{1}{2}x^2 + 5x - 4$

EasyMatch A.10(A)

STAAR – Algebra I: Quadratic and other nonlinear functions
The student will demonstrate an understanding of linear functions.

A.10 - Quadratic and other nonlinear functions. The student understands there is more than one way to solve a quadratic equation and solves them using appropriate methods. The student is expected to

A.10 (A) - Solve quadratic equations using [concrete] models, tables, graphs, and algebraic methods; **Readiness Standard**

STANDARD AND LESSON OVERVIEW
TEACHER DIRECTIONS

A.10 (A) - Solve quadratic equations using [concrete] models, tables, graphs, and algebraic methods; **Readiness Standard**

Tables and Graphs

Tables can be very useful in graphing quadratic functions. Make a table for the values which satisfy the quadratic function:

For example: $y = ax^2$, where $|a| > 1$, make a table of the values which satisfy the equation:

$y = 2x^2$

x	-1.5	-1	0	1	1.5
y	4.5	2	0	2	4.5

Plot the points from the table and draw a smooth curve through the points.

See Figure A.10.1 which follows.

KEEPING IT REAL - ALGEBRA I   A.10(A)

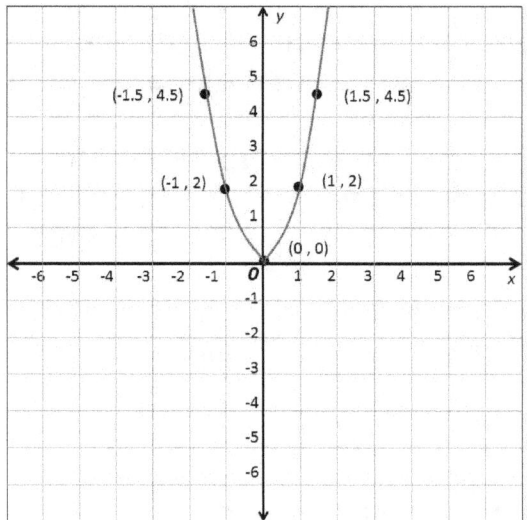

Figure A.10.1

Figure A.10.1 is a graph of the points of the table of points for **y = 2x²**. From the graph one can determine that the **domain** is all real numbers and the **range** is all y ≥ 0.

Properties of the Graph of a Quadratic Function

Recall the graph of y = ax² + bx + c is a parabola with the following properties:

1. Opens up if a > 0 and opens down if a < 0.

2. Is narrower than the graph of y = x² if |a| > 1 and wider if the |a| < 1.

3. Has an axis of symmetry of x = $x = \dfrac{-b}{2a}$

4. Has a vertex with an x-coordinate of $\dfrac{-b}{2a}$

5. Has a y-intercept of c, so the point (0, c) is on the parabola

Thus, to find the axis of symmetry and the vertex for the function y = -2x² + 12x – 6, one can use the following steps:

    y = -2x² + 12x – 6   **Notes**

    y = -2x² + 12x – 6   From the equation: a = -2 and b = 12.

KEEPING IT REAL - ALGEBRA I A.10(A)

$x = \dfrac{-b}{2a} = -\dfrac{12}{2(-2)} = 3$ Substituting the -2 for a and the 12 for b, and simplify

The x component is the axis of symmetry: x = 3

f(3) = -2(3)² + 12(3) – 6   To find the y-coordinate of the vertex

f(3) = -18 + 36 – 6   Substitute 3, for x

f(3) = -24 + 36

f(3) = 12   **The vertex is the point (3, 12)**

With a y-intercept of (0,-6), vertex of (3, 12) and axis of symmetry at x = 3, one can graph the function y = -2x² + 12x – 6, since one side of the parabola can be used to graph the other side.

   INTEGRATING TECHNOLOGY      CULTURAL COMPETENCY

   PROJECT-BASED LEARNING      WRITING ACROSS THE CURRICULUM

www.keepingitrealeasymatch.com

KEEPING IT REAL - ALGEBRA I

A.10(A)

Student Materials

Objectives for A.10 (A): Solving Quadratic Functions using models, tables, graphs and algebraic methods.

1. Factor the following expressions

   a. $x^2 - 64$

   b. $a^2 - 16$

   c. $x^2 - 6x + 9$

   d. $4x^2 + 12x + 9$

   e. $x^2 - 8x + 16$

2. Solve the following quadratic equations and graph.

   a. $x^2 - 2x - 3 = 0$ b. $x^2 - 4x - 5 = 0$ c. $x^2 - 4x + 5 = 0$

KEEPING IT REAL - ALGEBRA I   A.10(A)

3. Find the value of c that makes the expression a perfect trinomial square. Then write the expression as the square of a binomial.

   a. $x^2 + 8x + c$ b. $x^2 - 12x + c$ c. $x^2 + 3x + c$

4. Make tables for the values of the functions below. Graph the functions.

   a. $y = 3x^2$ b. $y = \frac{1}{2}x^2$ c. $y = -2x^2$

KEEPING IT REAL - ALGEBRA I

KEEPING IT REAL - ALGEBRA I

A.10(A)

5. For the quadratic equation $x^2 - 9 = 0$, what is the solution set?

6. Graph the function $y = x^2 + x - 6$. What are the values for x, when $x^2 + x - 6 = -4$?

EasyMatch A.10(B)

STAAR – Algebra I: Quadratic and other Nonlinear functions
The student will demonstrate an understanding of linear functions.

A.10 - Quadratic and other nonlinear functions. The student understands there is more than one way to solve a quadratic equation and solves them using appropriate methods. The student is expected to

A.10 (B) - Make connections among the solutions (roots) of quadratic equations, the zeros of their related functions, and the horizontal intercepts (x-intercepts) of the graph of the function. **Supporting Standard**

STANDARD AND LESSON OVERVIEW
TEACHER DIRECTIONS

A.10 (B) - Make connections among the solutions (roots) of quadratic equations, the zeros of their related functions, and the horizontal intercepts (x-intercepts) of the graph of the function. **Supporting Standard**

Zero-Product Property

The Zero-Product Property is a key concept in solving polynomial equations in factored form. Let **a** and **b** be real numbers. If **ab** = 0, then **a** = 0 or **b** = 0. The Zero-Product Property is used to solve an equation when one side is zero and the other side is a product of polynomial factors. **The solutions of such an equation are also called the roots**. Consider the quadratic function, $x^2 - 2x - 8 = 0$.

	Notes
$x^2 - 2x - 8 = 0$	
$(x - 4)(x + 2) = 0$	Factoring the original equation, $x^2 - 2x - 8 = 0$.
$x - 4 = 0$ or $x + 2 = 0$	Zero-product property
$x = 4$ or $x = -2$	Additive Inverse for both values
	The solutions of the equation are 4 and -2
	The solutions of 4 and -2 are also the roots of the equation

KEEPING IT REAL - ALGEBRA I   A.10(B)

Complex Numbers as Roots

Given $x^2 + 2x + 3 = 0$, solve and graph the solution.

$x^2 + 2x + 3 = 0$	**Notes**
$x^2 + 2x = -3$	Additive Inverse, -3 from both sides. Coefficient of x^2 is one (1).
$x^2 + 2x + 1 = -3 + 1$	**Completing the square: Take half of the coefficient of x, 2, square it and add to both sides. Half of 2 is one (1).**
$(x + 1)^2 = -2$	Writing the **Square of a binomial**
$x + 1 = \pm\sqrt{-2}$	Take the square root of both sides
$x + 1 = \pm\sqrt{2}\,i$	i^2 is defined to be a negative one (-1)
$x = -1 \pm \sqrt{2}\,i$	Notes: $\sqrt{-4} = \pm 2i$, then $(2i)(2i) = 4i^2$ and $4(-1) = -4$ and $(-2i)(-2i) = 4i^2 = -4$ Since the roots are complex numbers, the graph **does not cross the x-axis. Thus, there is no solution.**
$x = \dfrac{-b}{2a} = \dfrac{-2}{2(1)} = -1$	x-component of vertex is -1. $f(x) = x^2 + 2x + 3$
$f(-1) = -1^2 + 2(-1) + 3$	Substituting -1 for x, to find f(x)
$f(-1) = 1 - 2 + 3 = 2$	**y-component of vertex, thus vertex is (-1, 2).**

The graph for the equation, $x^2 + 2x + 3 = 0$, is represented in Figure A.10.2 below.

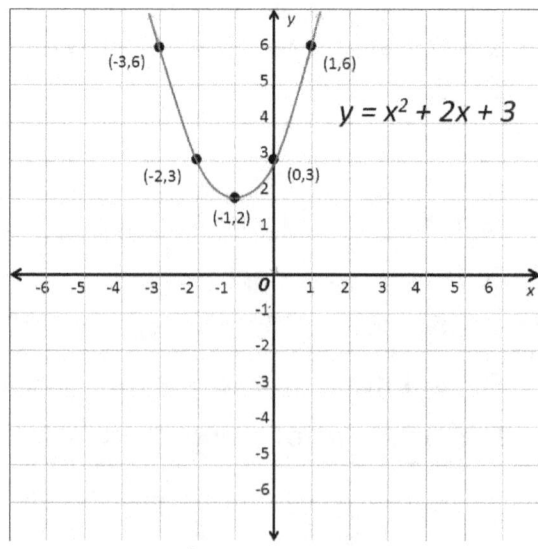

Figure A.10.2

KEEPING IT REAL - ALGEBRA I A.10(B)

Substituting the selected values for x, the y-values are determined. Going back to the equation where $(x + 1)^2 = -2$:

$(x + 1)^2 = -2$	**Notes**
$(x + 1)^2 + 2 = 0$	Additive Inverse, adding 2 to both sides.
$f(-1) = (x+1)^2 + 2$	Since $x = -1$,
$f(-1) = (-1+1)^2 + 2$	substituting a -1 for x,
$f(-1) = 2$	**Thus the vertex is (-1, 2)**

Note: The domain is all real numbers and the range is $y \geq 2$. The graph does not cross the x-axis, thus there is no solution.

Quadratic Formula

By completing the square for the quadratic equation $ax^2 + bx + c = 0$, one can develop a formula that gives the solutions of any quadratic equation in standard form. This formula is called the quadratic formula. The quadratic formula was derived in section A.9(A) under the **Completing the Square** section and is provided below:

$$x = \frac{-b \pm \sqrt{b^2 - 4ac}}{2a}$$

The quadratic formula will be used to solve for $x^2 - 6x + 3 = 0$.

For $x^2 - 6x + 3 = 0$, $a = 1$, $b = -6$ and $c = 3$, substituting into the quadratic formula:

	Notes
$x = \dfrac{-(-6) \pm \sqrt{(-6)^2 - 4(1)(3)}}{2(1)}$	Substituting the values for a, b and c
$x = \dfrac{6 \pm \sqrt{36 - 12}}{2}$	Performing operations
$x = \dfrac{6 \pm \sqrt{24}}{2}$	Simplify
$x = \dfrac{6 \pm \sqrt{4 \times 6}}{2}$	Product property of radicals
$x = \dfrac{6 \pm 2\sqrt{6}}{2}$	Property of radicals
$x = 3 \pm \sqrt{6}$	Simplify

$x = 3 \pm \sqrt{6}$ and $x = 3 - \sqrt{6}$ The solutions to the equation are:

$$x = 3 + \sqrt{6} \text{ and } x = 3 - \sqrt{6}$$

Solve Quadratic Equations by Graphing

For a quadratic equation written in the standard form, one can also use graphing to solve the equation. As in the example above, **the solutions or roots** to the quadratic equation are the x-intercepts of the graph.

Given the quadratic equation: **x² - 2x - 3 = 0**

Notes

x² - 2x - 3 = 0	**y = ax² + bx + c**, is the standard form
(x-3) (x+1) = 0	**Factoring, writing as product of other polynomials**
x−3=0 x+1=0	Using the zero-product property, set both results equal to 0
x = 3 x = -1	**Thus, the graph will cross the x axis at points (3, 0) and (-1, 0)**

Figure A.9.3 is provided below.

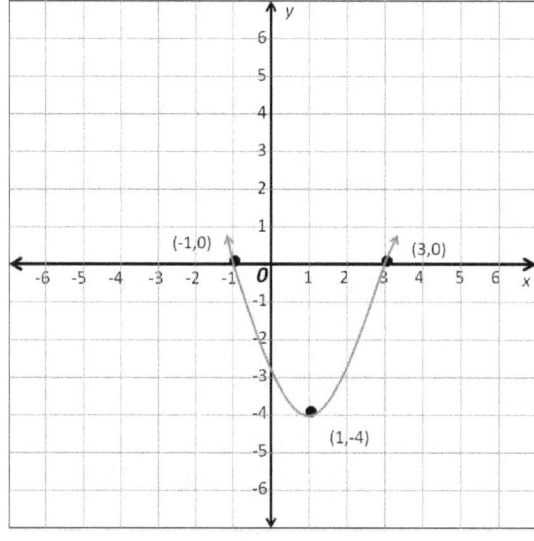

Figure A.9.3

KEEPING IT REAL - ALGEBRA I   A.10(B)

The standard approach to find the vertex can be used to complete the graph.

$x = \dfrac{-(-2)}{2(1)}$ To find the vertex, **the x component of the vertex is:** $x = \dfrac{-b}{2a}$

x = 1   Using f(x) = x² - 2x - 3, **to find the y component of the vertex**

f(1) = (1)² -2(1) - 3   **by substituting the value of x, to find f of x, or f of (x = 1)**

f(1) = 1 − 2 − 3

f(1) = -4   **The y component is -4, thus the Vertex is (1, -4).**

Note: For graphs with no x-intercepts, the equation has no solution.

Zeros of a Quadratic Function

The zero of a function is identified as an x-intercept of the function's graph. It is common to use the function's graph to find the zeros of the function. For the graph in Figure A.9.3 above, **the x-intercepts are 3 and -1, thus the zeros of the function are 3 and -1.**

Approximating the Zeros of a Function

Using a table of values to estimate the zeros of a function requires one to look for the change in signs of the function **(y)** values. For example, for the functions, $y = \dfrac{1}{2}x^2 - 4$ and $y = x^2$ as illustrated in Figure A.9.4 below, please note the table of values for $y = \dfrac{1}{2}x^2 - 4$. In the table, the signs for the function values of **y** change from positive to negative between x = -4 and -2 and the signs for **y** change again from negative to positive between the values of x = 2 and x = 4. One could establish a table of values for the range of x to be between -4 and -2 and also for the values for x between 2 and 4. The x values closest to the sign changes in the function **(y)** values are the approximate zeros of the function.

KEEPING IT REAL - ALGEBRA I   A.10(B)

Note the comparison of the graphs for the two functions below.

$y = \frac{1}{2}x^2 - 4$ and $y = x^2$ See Figure A.9.4.

Make a table of values for $y = \frac{1}{2}x^2 - 4$.

x	-4	-2	0	2	4
y	4	-2	-4	-2	-4

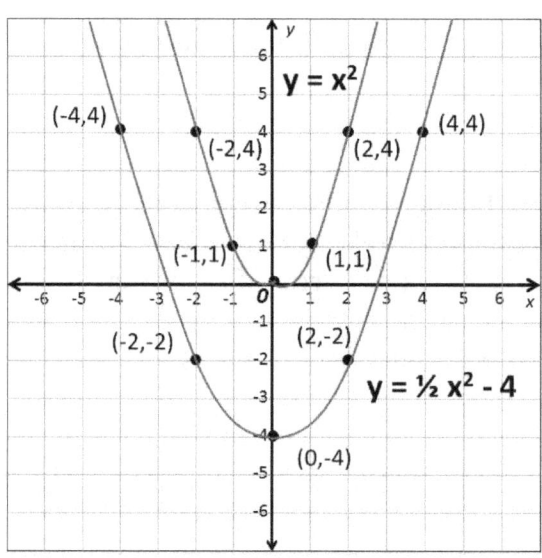

Figure A.9.4

 INTEGRATING TECHNOLOGY

 CULTURAL COMPETENCY

 PROJECT-BASED LEARNING

 WRITING ACROSS THE CURRICULUM

www.keepingitrealeasymatch.com

KEEPING IT REAL - ALGEBRA I   A.10(B)

Student Materials

Objectives for A.10 (B): Making connections among the solutions (roots) of quadratic equations, the zeros of their related functions, and the horizontal intercepts (x-intercepts) of the graph.

1. Use the quadratic formula to solve the following equations.

   a. $x^2 - 5x = 14$

   b. $x^2 + 5 - 10 = 0$

   c. $-z^2 + z + 14 = 0$

   d. $-2n^2 - 5n + 16 = 0$

   e. $3x^2 - 4 = 11x$

2. Solve the following quadratic equations by graphing.

   a. $x^2 + 5x + 6 = 0$

   b. $x^2 - 4x = 5$

   c. $x^2 - 6x + 9 = 0$

   d. $x^2 + x = -2$

   e. $\frac{1}{2}x^2 + 2x = 6$

KEEPING IT REAL - ALGEBRA I   A.10(B)

3. Find the zeros of the equations for the quadratic functions.

   a. $y = x^2 + 3x - 10$

   b. $y = -x^2 + 8x + 9$

   c. $y = x^2 - 5x - 6$

   d. $y = x^2 - 12x + 11$

   e. $y = -x^2 - 7x + 8$

4. The table below represents values for the quadratic function h. If 3 is one solution for h(x) = 0, what is the value for the other solution?

x	h(x)
-8	-2.75
-7	0
-6	2.25
-5	5.25

5. For the quadratic function $f(x) = 5x^2 + 4x - 1$, what are the x-intercepts of the graph?

> # EasyMatch A.11(A)
>
> ### STAAR – Algebra I: Linear Functions
> **The student will demonstrate an understanding of linear functions.**
>
> **A.11 - Quadratic and other nonlinear functions.** The student understands there are situations modeled by functions that are neither linear nor quadratic and models the situations. The student is expected to
>
> A.11 (A) - Use patterns to generate the laws of exponents and apply them in problem-solving situations; **Supporting Standard**

STANDARD AND LESSON OVERVIEW
TEACHER DIRECTIONS

A.11 (A) - Use patterns to generate the laws of exponents and apply them in problem-solving situations; **Supporting Standard**

Exponents

Definitions: A **power** is an expression that represents repeated multiplication of the same factor. For example, 81 is a power of 3 because $81 = 3 \times 3 \times 3 \times 3$. A power can be written in a form using two numbers, a base and an exponent. The **exponent** represents the number of times the **base** is used as a factor, so 81 can be written as 3^4, where 3 is the base and 4 is the exponent. x^5 means that x is the base and 5 is the exponent.

Expanding $x^5 = x \times x \times x \times x \times x$. For $x^5 \times x^3 = x \times x \times x \times x \times x \times x \times x \times x$. There are 8 x's which implies that $x^5 \times x^3 = x^8$. So, in multiplying exponents with the same base, the exponents are added such that $x^n \times x^m = x^{n+m}$. In dividing with the same base,

$\dfrac{x^5}{x^3} = x^{5-3}$ therefore, $\dfrac{x^n}{x^m} = x^{n-m}$ Example: $\dfrac{x \times \cancel{x} \times \cancel{x} \times \cancel{x} \times x \times x}{\cancel{x} \times \cancel{x} \times \cancel{x}} = x^2$

For $(x^2)^3$, implies that x^2 is multiplied by itself 3 times illustrated as $x^2 \times x^2 \times x^2 = x^{2+2+2} = x^6$.

Therefore $(x^n)^m = x^{n \times m}$

Note: Any number divided by itself, other than zero (0), is equal to one (1). Since $1 = \dfrac{x^n}{x^n}$, which can be written as $x^{n-n} = x^0$, therefore any x^0 power is equal to one (1).

203

KEEPING IT REAL - ALGEBRA I   A.11(A)

Given a number x^n and multiplying x^n by one (1) expressed as $\frac{x^n}{x^n}$,

hence $x^{-n} = x^{-n} \times \frac{x^n}{x^n} = \frac{x^0}{x^n}$. Since $x^0 = 1$, then $x^{-n} = \frac{1}{x^n}$.

Given $\frac{1}{x^{-n}}$ and multiplying $\frac{1}{x^{-n}}$ by one in the form of $\frac{x^n}{x^n}$ gives, $\frac{1}{x^{-n}} = \frac{1}{x^{-n}} \times \frac{x^n}{x^n} = \frac{x^n}{x^0}$. Since any number to the zero (0) power is one, then $\frac{1}{x^{-n}} = \frac{x^n}{x^0} = \frac{x^n}{1} = x^n$. Therefore, $\frac{1}{x^{-n}} = x^n$.

Given $\frac{1}{x^n}$ multiplied by one (1) expressed as $\frac{x^{-n}}{x^{-n}}$ then $\frac{1}{x^n} = \frac{1}{x^n} \times \frac{x^{-n}}{x^{-n}} = \frac{x^{-n}}{x^0} = \frac{x^{-n}}{1} = x^{-n}$.

Therefore, $\frac{1}{x^n} = x^{-n}$.

If given $\sqrt[n]{x^m} = x^{m/n}$, take the n root of the number and raise that number to the m power.

If there is no n in the index, then the square root is implied.

Example: $\sqrt[3]{8^4} = x^{4/3}$ which implies to take the cube root of 8 and raise it to the fourth power.

Must Know the Rules and Principles of Exponents:

$$x^n \times x^m = x^{n+m}$$

$$\frac{x^n}{x^m} = x^{n-m}$$

$$(x^n)^m = x^{n \times m}$$

$$x^0 = 1$$

$$x^{-n} = \frac{1}{x^n}$$

$$x^n = \frac{1}{x^{-n}}$$

$$\sqrt[n]{x^m} = x^{m/n}$$

Examples in reading and writing powers:

Expression	Notes
6^1	Six (6) to the first power = 6
4^2	Four (4) to the second power or four squared: 4 x 4 = 16
$(\frac{1}{2})^3$	$\frac{1}{2}$ to the third power or $\frac{1}{2} \times \frac{1}{2} \times \frac{1}{2} = \frac{1}{8}$
d^4	d to the fourth power or: d x d x d x d

KEEPING IT REAL - ALGEBRA I — A.11(A)

 INTEGRATING TECHNOLOGY

 CULTURAL COMPETENCY

 PROJECT-BASED LEARNING

 WRITING ACROSS THE CURRICULUM

www.keepingitrealeasymatch.com

KEEPING IT REAL – ALGEBRA I A.11(A)

Student Materials

Objectives for A.11 (A): Use patterns to generate the laws of exponents and apply them in problem-solving situations

1. Simplify the following expressions and provide the product:

   a. $4^2 \times 4^3$

   b. $(7^2)^3$

   c. $n^3 \times n^2$

   d. $(-2)^3$

   e. 5×5^4

2. Simplify the expression and provide your answer.

   a. $(-3)^2 \times 5$

   b. $\dfrac{x^5}{x^3}$

   c. $\dfrac{2^3}{2^2}$

   d. $\dfrac{-3^5}{-3^3}$

   e. $\dfrac{4^7}{4^4}$

KEEPING IT REAL - ALGEBRA I   A.11(A)

3. Find an equivalent expression for: $\dfrac{8x^6 y^{-4}}{3x^2 y^{-6}}$

4. Find an equivalent expression for: $\dfrac{12x^6 y^2 z^4}{3x^2 y^{-2} z^3}$

5. Find an equivalent expression for: $\dfrac{6x^6 2y^{-4} z^2}{3x^2 2y^{-4} z^{-3}}$

KEEPING IT REAL - ALGEBRA I

A.11(A)

EasyMatch A.11(B)

STAAR – Algebra I: Linear Functions
The student will demonstrate an understanding of linear functions.

A.11 - Quadratic and other nonlinear functions. The student understands there are situations modeled by functions that are neither linear nor quadratic and models the situations. The student is expected to

A.11 (B) - Analyze data and represent situations involving inverse variation using [concrete] models, tables, graphs, or algebraic methods; **Supporting Standard**

STANDARD AND LESSON OVERVIEW
TEACHER DIRECTIONS

A.11 (B) - Analyze data and represent situations involving inverse variation using [concrete] models, tables, graphs, or algebraic methods; **Supporting Standard**

Considering the two variables x and y, **direct variation is expressed by, y = ax and a ≠ 0.** The variables x and y express **inverse variation if** $y = \dfrac{a}{x}$ **and a ≠ 0.** The nonzero number **a** is the constant of variation, and **y varies inversely with x.**

Direct Variation (y = ax and a ≠ 0) and Inverse Variation ($y = \dfrac{a}{x}$ and a ≠ 0)

	Examples	Notes
1.	xy = 6	Original equation
	$y = \dfrac{6}{x}$	Multiplicative Inverse, multiply both sides by $\dfrac{1}{x}$
		Because xy = 6 can be written in the form $y = \dfrac{a}{x}$, xy = 6 represents **inverse variation**. The constant of variation is 6.

KEEPING IT REAL - ALGEBRA I   A.11(B)

2. $\dfrac{y}{4} = x$ Original equation

   $y = 4x$ Multiplicative Inverse, multiply both sides by 4

   Because $\dfrac{y}{4} = x$ can be written in the form $y = ax$, $\dfrac{y}{4} = x$ represents **direct variation.**

3. $y = 2x + 5$ Original equation

   Because $y = 2x + 5$ cannot be written in the form of $y = ax$ or $y = \dfrac{a}{x}$, $y = 2x + 5$ does not represent either direct variation or inverse variation.

Graphs of Direct Variation and Indirect Variation

The graphs of direct variation and indirect variation are illustrated in Figures A.11.1 and A.11.2 below.

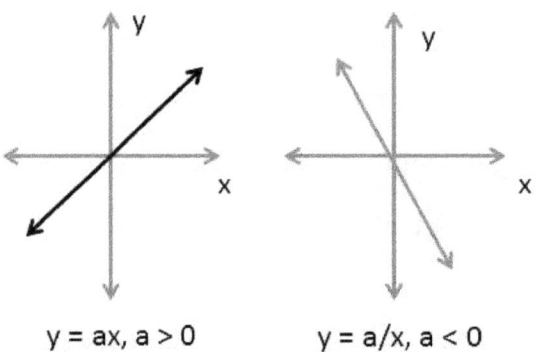

y = ax, a > 0   y = a/x, a < 0

Figure A.11.1 - Direct Variation

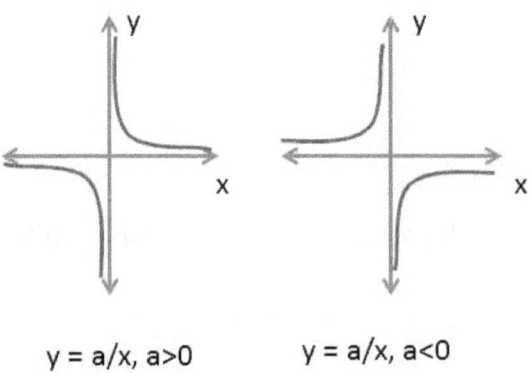

y = a/x, a > 0   y = a/x, a < 0

Figure A.11.2 - Inverse Variation

KEEPING IT REAL - ALGEBRA I   A.11(B)

For inverse variation, there are two symmetrical parts of a hyperbola called branches of the hyperbola. The lines that the hyperbola approaches but does not interest are called the asymptotes of the hyperbola. The asymptotes of the graph of the $y = \dfrac{a}{x}$, are the x-axis and the y-axis.

As an example, the graph of the inverse variation equation $y = \dfrac{6}{x}$, can be accomplished with the aid of a table of values.

To develop a table for the equation, $y = \dfrac{6}{x}$, choose several integer values for x and find the values of y. Both positive and negative values are chosen to fit the scale of the graph.

x	1	1.5	2	3	4	6
y	6	4	3	2	1.5	1

x	-1	-1.5	-2	-3	-4	-6
y	-6	-4	-3	-2	-1.5	-1

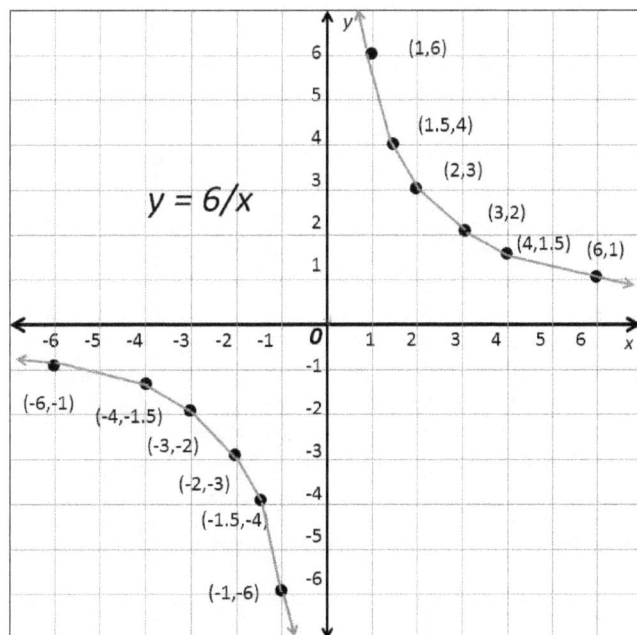

Figure A.11.3

Figure A.11.3 illustrates the graph for equation, $y = \dfrac{6}{x}$.

KEEPING IT REAL - ALGEBRA I A.11(B)

 INTEGRATING TECHNOLOGY

 CULTURAL COMPETENCY

 PROJECT-BASED LEARNING

 WRITING ACROSS THE CURRICULUM

www.keepingitrealeasymatch.com

KEEPING IT REAL - ALGEBRA I   A.11(B)

Student Materials

Objectives for A.11 (B): Analyze data and represent situations involving inverse variation using [concrete] models, tables, graphs, or algebraic methods;

1. Determine whether the equation represents direct variation, inverse variation or neither.

   a. $y = -3x$ b. $xy = 2$

   c. $\dfrac{y}{x} = 5$ d. $2x + y = 4$

   e. $x = \dfrac{-1}{y}$

2. Graph the inverse variation equations below.

   a. $y = \dfrac{-2}{x}$ b. $y = \dfrac{3}{x}$

   c. $y = \dfrac{4}{x}$ d. $y = \dfrac{5}{x}$

   e. $y = \dfrac{2}{x}$

KEEPING IT REAL - ALGEBRA I   A.11(B)

3. Graph the following and determine whether inverse variation.

a.

x	3	6	9	12	15	18
y	1	2	3	4	5	6

b.

x	-12	-10	-8	-5	-4	-2
y	2	2.4	3	4.8	6	8

EasyMatch A.11(C)

STAAR – Algebra I: Linear Functions
The student will demonstrate an understanding of linear functions.

A.11 - Quadratic and other nonlinear functions. The student understands there are situations modeled by functions that are neither linear nor quadratic and models the situations. The student is expected to

A.11 (C) - Analyze data and represent situations involving exponential growth and decay using [concrete] models, tables, graphs, or algebraic methods.
Supporting Standard

STANDARD AND LESSON OVERVIEW
TEACHER DIRECTIONS

A.11 (C) - Analyze data and represent situations involving exponential growth and decay using [concrete] models, tables, graphs, or algebraic methods.
Supporting Standard

The exponential function has the form **y = ab^x where a ≠ 0, b > 0 and b ≠ 1**. The exponential function is nonlinear. Comparing a linear function to an exponential function, one should notice the primary difference in the equations:

Linear Function: y = 2x + 3

Exponential Function: y = 3 x 2^x

Analyzing the data for the two functions, one can notice the difference.

Linear Function: **y = 2x + 3**

x	-3	-2	-1	0	1	2	3
y	-3	-1	1	3	5	7	9

Note: Each increase is plus 2 or + 2

Exponential Function: **y = 3 x 2^x**

x	-3	-2	-1	0	1	2	3
y	$\frac{3}{8}$	$\frac{3}{4}$	$\frac{3}{2}$	3	6	12	24

Note: Each increase is times 2 or x2

KEEPING IT REAL - ALGEBRA I   A.11(C)

Writing the rule for the exponential function, **y = 3 x 2^x**, the **y** values are multiplied by 2 for each increase of 1 in **x**, so the exponential function is of the form **y = ab^x, where b = 2**. To find the value of **a**, find the value of **y** when **x = 0**, so **a = 3**.

Graphing Exponential Functions

Graphing an exponential function is often done by making a table and choosing values for x and finding the values for y. The graph the exponential function **y = 2^x**, make at table with values, plot the points and draw a smooth curve through the points.

$$y = 2^x$$

x	-2	-1	0	1	2
y	$\frac{1}{4}$	$\frac{1}{2}$	1	2	4

The graph for the exponential function **y = 2^x** is illustrated below in Figure A.11.4.

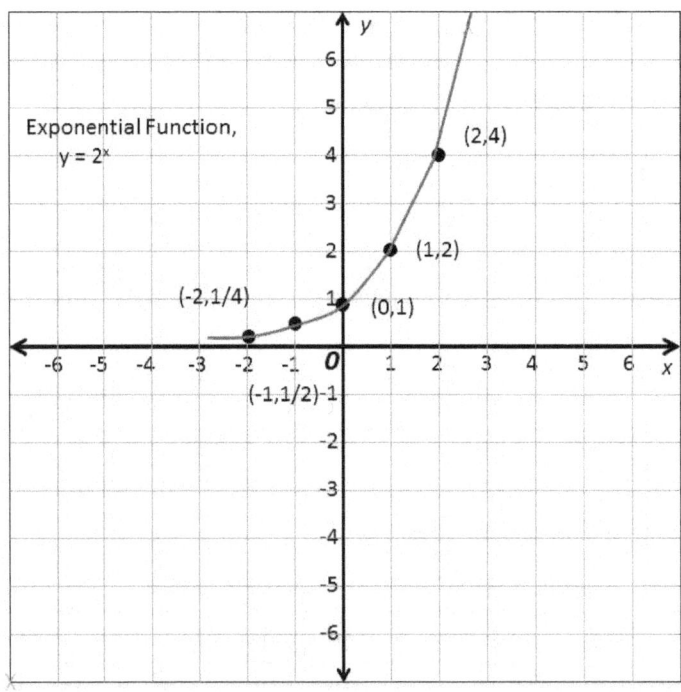

Figure A.11.4

KEEPING IT REAL - ALGEBRA I   A.11(C)

Exponential Growth and Decay

When a > 0 and b > 1, the exponential function **y = ab^x** illustrates **exponential growth**. For exponential growth, the quantity increases by the same percentage over equal time periods. To model exponential growth, the variables are:

For **y = ab^x**   **a** is the initial amount;

b is the growth factor and

x is the number of increases

Another approach:

For **y = a(1 + r)^t**   **a** is the initial amount;

(1 + r) is the growth factor, where **r** is the growth rate

t is the time period

Figure A.11.4 exemplifies the **exponential growth** function where **y = ab^x**, where a = 1.

An example for **y = a(1 + r)^t**, can be exemplified for compound interest. Compound Interest is interest earned on both an initial investment and on previously earned interest. Compound of interest can be modeled by exponential growth where **a** is the initial investment, **r** is the annual interest rate, and **t** is the number of years the money is invested.

Suppose you put $500 in a savings account that earns 4% annual interest compounded annually. If there are no additional deposits or withdrawals, how much will your investment be worth in 5 years?

Notes

y = a(1 + r)^t	Exponential growth model
y = 500(1 + 0.04)^5	Substituting, 500 for a, 0.04 for r and 5 for t
y = 500(1.04)^5	Simplify
y = 608.33	Your investment will total $608.33 in 5 years

KEEPING IT REAL - ALGEBRA I   A.11(C)

Exponential Decay

When a > 0 and 0 < b < 1, the function **y = ab^x** exemplifies **exponential decay**. The graph of the exponential growth function rises from left to right, and the graph of the exponential decay function falls from left to right.

For $y = a(1 - r)^t$ *a* is the initial amount;

                       (1 - r) is the decay factor, where *r* is the decay rate

                       t is the time period

Example: Small farmers decreased in the southern United States from 1960 to 2010 by 2% annually. In 1960 there were approximately 40 thousand small farms in the southern United States. Write a function which models the number of small farmers in southern United States over time and approximate how many small farmers there were in 2010.

$F = a(1 - r)^t$ Exponential decay model

$F = 40(1 - 0.02)^{50}$ Substitute 40 for a, and 0.02 for r

$F = 40(.98)^{50}$ Simplify

$F = 14.57$ There were approximately 14,568 small farmers in 2010.

 INTEGRATING TECHNOLOGY    CULTURAL COMPETENCY

 PROJECT-BASED LEARNING    WRITING ACROSS THE CURRICULUM

www.keepingitrealeasymatch.com

KEEPING IT REAL - ALGEBRA I   A.11(C)

Student Materials

Objectives for A.11 (C): Analyze data and represent situations involving exponential growth and decay using [concrete] models, tables, graphs, or algebraic methods;

1. Tell whether the table represents an exponential function.

x	-4	-3	-2	-1	0	1	2	3	4
y	$\frac{3}{16}$	$\frac{3}{8}$	$\frac{3}{4}$	$\frac{3}{2}$	3	6	12	24	48

_____

_____

2. Graph the following functions and determine if they demonstrate exponential growth or decay.

    a. $y = (0.3)^x$

    b. $y = 5(0.2)^x$

    c. $y = 3^x$

    d. $y = 4 \times 2^x$

    e. $y = 8 \times 3^x$

KEEPING IT REAL - ALGEBRA I   A.11(C)

3. You purchased a cell phone for $250. The value of the phone decreases by 30% annually. Write a function that models the value of your phone over time. What is the value of your phone after 2 years?

_____

_____

_____

4. The table below provides the values for the exponential function h. How would one describe the function?

x	h(x)
1	140,000
2	143,850
3	147,806
4	151,871

_____

_____

_____

5. Suppose your parents deposited $10,000 in an account paying 7% interest, compounded annually when you were born. The funds were to provide a new car upon high school graduation. At the age of 18, how much money will you have for a car?

_____

_____

_____

ABOUT THE AUTHORS

Dr. Patricia Hoffman-Miller

Dr. Hoffman-Miller is an experienced executive with over thirty years of management and executive experience in private, post-secondary and P-20 organizations. As a private sector and school district executive, she managed and administered Human Resources and Labor Relations departments for Rockwell International, The Pillsbury Corporation, Ross Electronics, and the Gary Community School Corporation, negotiating labor contracts with five major bargaining units. She developed the first executive and professional leadership program in concert with local universities in Cedar Rapids and Iowa City, Iowa.

As an Assistant Professor of Management and a Primary Professor, Dr. Hoffman-Miller taught undergraduate and graduate students in the following disciplines: management and organizational theory, financial administration, business communications, accounting, labor relations, ethical decision making and labor economics.

Dr. Hoffman-Miller has over 13 years of experience in P-12 public and public charter schools. She held positions as a classroom teacher, Assistant Principal, Principal and Assistant Superintendent in school districts in the eastern and mid-western sections of the country. She developed the first state certified mentor program for teachers in a mid-western school district. Under her leadership, a principal efficacy program (E-PLUS: Enhancing Principal Leadership in Urban Schools) was designed to assist struggling building administrators become more effective advocates for student centered learning.

After success as a traditional school and district administrator, Dr. Hoffman-Miller became actively involved in the charter school movement. In her capacity as principal, regional vice president and national assessment director, Dr. Hoffman-Miller was responsible for seven successful start-up schools, managing and administering all phases of new charter school start-ups including, but not limited to: operational procedures, inclusive of manuals, fiscal/budgetary processes, staff and administrator recruitment and selection, facilities management, and curricula implementation, alignment and assessment.

Dr. Hoffman-Miler simultaneously improved student performance in Pennsylvania and the District of Columbia. Howard Road Academy, a District of Columbia Charter School, was facing closure in September, 2005 when Dr. Hoffman-Miller was appointed Regional Vice President. After one year in implementing Dr. Hoffman-Miller's prescriptive approach, Howard Road Academy made AYP using safe Harbor. During the second year of her leadership, Howard Road Academy became one of the top performing schools in the District of Columbia.

Dr. Hoffman-Miller has traveled extensively across the country working with charter schools in Indiana, Colorado, Arizona, Ohio, Pennsylvania, Louisiana and the District of Columbia. She directly managed all operational, fiscal and academic programs in three states, with school and student performance increasing exponentially at each site. All schools under her leadership made AYP during the first year.

As Interim Department Chair, Department of Kinesiology and Allied Health Science, Dr.

Hoffman Miller was able to motivate faculty members to approve a required Departmental name change from Health and Human Performance to Kinesiology and Allied Health Sciences. The proposed change was approved by the University Academic Council. She assumed Leadership of the Department on November 1, 2008.

She served the Whitlowe R. Green College of Education in a number of administrative capacities including Interim Department Chair, Curriculum and Instruction and Interim Associate Dean, She currently serves as Associate Professor in the Whitlowe R. Green College of Education where she successfully led the SACS accreditation process and spearheaded national recognition of the building principal preparation program by ELCC in 2010 and again in 2012. She teaches in the doctoral program in the Department of Educational Leadership and Counseling and is responsible for chairing dissertation committees of five doctoral candidates.

Dr. Freddie L. Frazier

Dr. Frazier is a Distinguished Professor of Mathematics at Prairie View A&M University. His experiences as an instructor and professor of mathematics extend beyond fifty years. During his five year tenure in public schools, Dr. Frazier served as a mathematics teacher, department head and taught night classes for adults. Subsequently, Dr. Frazier transitioned to Prairie View A&M University and shortly thereafter began a lifelong tenure as the coordinator of engineering mathematics.

Dr. Frazier is the founder of Frazier Educational Enterprises, Inc., providing educational products, services and consulting. He developed complete courses in Basic Mathematics, Algebra, Trigonometry, Differential Calculus and Differential Equations, which are available on DVD. Each course is presented as a set of six (6) or eight (8) DVD's, which allows easy and convenient access to Dr. Frazier's proposed course content and teaching methods.

As an educational consultant, Dr. Frazier has provided lectures and services to over thirty schools and school districts in efforts to improve student and teacher performance in mathematics. It is well documented that Dr. Frazier has been instrumental in enhancing public, chartered and alternative schools and school districts from poor performance to exceptional performance. His approaches to teacher training and motivating students are unique and serve as models to help enhance academic performance.

Dr. Frazier's philosophy of teaching mathematics is founded on a basic concept: "If you can count and know your left hand from your right hand, you can learn mathematics." The simple concept of counting is used throughout basic mathematics and helps to build student confidence while learning. He developed the Embedded Spiraling Bond (ESB) approach to teaching mathematics which exemplifies the use of the Additive Inverse, Multiplicative Inverse and Distributive Property in Basic Mathematics and Algebra. The ESB method has proven success in student learning and performance at the middle school, high school and collegiate levels.

Dr. Freddie Frazier worked very hard to establish and maintain the STEM Enhancement Program, sponsored by the National Science Foundation (NSF) at Prairie View A&M University. The ten year, $6.3 million program provided scholarships for over 650 students. The overall cumulative GPA for the students graduating from the program was above a 3.0/4.0 for the 10-year period, with a graduation rate of 70 per cent. The program served as a model for the more than fifty (50) universities with similar NSF programs. In addition, the STEM Enhancement Program was used as a model by the Houston Area Urban League to establish local STEM Programs for Pre-School, Middle School and High School Students.

Dr. Frazier is the recipient of numerous awards for teaching mathematics and community service at the national, departmental and local levels. He has been very successful in acquiring research and scholarship funding from the Central Intelligence Agency, National Science Foundation and United States Navy. Over the recent twenty years, Dr. Frazier was instrumental in acquiring over $12 million for research and education.

Dr. Kelvin Kirby

Dr. Kirby is an Associate Professor of Electrical and Computer Engineering at Prairie View A&M University. His primary area of research and expertise is STEM Education. He has organized teams to access funding for STEM education and research from the National Science Foundation, National Aeronautics and Space Administration and the Department of Education,

Dr. Kirby has received numerous awards for teaching, research and service from Lockheed Martin, Prairie View A&M National Alumni Association, National Society of Black Engineers, Institute of Electrical and Electronic Engineers and several local organizations.

He worked with Dr. Freddie Frazier to establish the STEM Enhancement Program, sponsored by the National Science Foundation at Prairie View A&M University. The ten year, $6.3 million program provided scholarships for over 650 students. The overall cumulative GPA for the program participants was above a 3.0/4/0 for the 10-year period. In addition, the graduation rate for all participants was 70 per cent.

Dr. Kirby served on active duty in the United States Army for over thirteen years as a commissioned officer. During his military service, he gained extensive experience in training and skill development. His military experience coupled with the STEM Education allows him to be very effective in educational program development and execution. During his past twenty years at Prairie View A&M University, he has formed teams and served on teams which have acquired over $25 million dollars in STEM Education and research. He is an effective team builder and very skillful at maintaining program operations and achieving program goals.

Mrs. Frances Frazier

Mrs. Frazier is a Master Teacher with 46 years of experience teaching mathematics in the Houston Independent School District, in Houston, Texas. Mrs. Frazier has mentored and coached new secondary-level mathematics teachers and elementary certified teachers for over 20 years.

Mrs. Frazier's instructional expertise emphasizes developing students' critical thinking skills through mathematics problem solving. Over the years, she has identified numerous teaching strategies to assist students in translating word problems into numerical structured equations.

Though Mrs. Frazier in now retired, she still provides pro bono tutoring services in mathematics to a wide range of students, from the elementary level through helping to coach adults preparing to take professional teachers examinations for career advancement. In addition to her benevolence in tutoring youth and adults alike, Mrs. Frazier has demonstrated a heart to help educate all children.

As early as 2001, Mrs. Frazier began supporting an educational initiative to collect teaching supplies and educational materials for children in schools in The Gambia, West Africa. Within her capacity as President of Frazier Educational Enterprises, Inc., she procured technology (equipment) to accompany educational videos to be delivered to local schools, including a special donation to the teacher development certification program of Gambia College. The Gambia College donation was designed to aid the program in improving the mathematics competency of their teachers on both content knowledge and modeling of pedagogical techniques for teaching mathematics in elementary and secondary school classrooms.

Sales Order

Keeping it Real - EasyMatch

SHIP TO

NAME   NAME
ORGANIZATION   ORGANIZATION
ADDRESS   ADDRESS
CITY   CITY
STATE - ZIP   STATE - ZIP
TELEPHONE   TELEPHONE
PURCHASE ORDER #   EMAIL ADDRESS
CREDIT CARD #

P.O. Number	Shipping Method	Shipping Terms	Delivery Date	Payment Terms	Due Date
				Due on receipt	

Qty	Item #	Description	Unit Price	Discount	Line total
			Total Discount		
				Subtotal	
				Sales Tax	
				Total	

Make all checks payable to EDUCATIONAL CONCEPTS

Thank you for your business!

12320 Barker Cypress Road, Suite 600-111, Cypress, Texas 77429
P: 1-888-630-6650   F: 1-877-310-6692

www.ingramcontent.com/pod-product-compliance
Lightning Source LLC
Chambersburg PA
CBHW080733300426
44114CB00019B/2572